Springer Geography

The Springer Geography series seeks to publish a broad portfolio of scientific books, aiming at researchers, students, and everyone interested in geographical research. The series includes peer-reviewed monographs, edited volumes, textbooks, and conference proceedings. It covers the entire research area of geography including, but not limited to, Economic Geography, Physical Geography, Quantitative Geography, and Regional/Urban Planning.

More information about this series at http://www.springer.com/series/10180

Wade Bishop • Tony H. Grubesic

Geographic Information

Organization, Access, and Use

Wade Bishop
College of Communication and Information
University of Tennessee
Knoxville, TN, USA

Tony H. Grubesic
Center for Spatial Reasoning & Policy Analytics
College of Public Service & Community Solutions
Arizona State University
Phoenix, AZ, USA

ISSN 2194-315X ISSN 2194-3168 (electronic)
Springer Geography
ISBN 978-3-319-79428-0 ISBN 978-3-319-22789-4 (eBook)
DOI 10.1007/978-3-319-22789-4

Printed on acid-free paper

This Springer imprint is published by Springer Nature
The registered company is Springer International Publishing AG
The registered company address is: Gewerbestrasse 11, 6330 Cham, Switzerland

Wade Bishop would like to dedicate this book to four supportive and inspiring women, Jenn, Anna, Mary Lee, and Charlotte Bishop, and all the other residents of earth they share the world and its GI with and all the other residents of earth with whom they share this world and its geographic information.

Acknowledgments

The authors would like to acknowledge funding from the Institute of Museum and Library Services (IMLS). IMLS is the primary source of federal support for the nation's 123,000 libraries and museums. Through grant making, policy development, and research, IMLS helps communities and individuals thrive through broad public access to knowledge, cultural heritage, and lifelong learning. In 2012, the authors were awarded funding for the Geographic Information Librarianship (GIL) project from the Laura Bush 21st Century Librarian Program grant via the Institute of Museum and Library Services (IMLS) in their "Programs to Build Institutional Capacity" category.

The authors would also like to thank all the information professionals who participated in the Geographic Information Librarianship (GIL) project data collection, the ongoing efforts of the members of the American Library Association's Map and Geospatial Information Round Table (MAGIRT), and all the project's advisory committee—Michael Goodchild, *Emeritus Professor* at the University of California, Santa Barbara; Carol McAuliffe, *Head of the Map and Imagery Library*, University of Florida; Kathy Weimer, *Head, Kelley Center for Government Information, Data and Geospatial Services*, Rice University Libraries; Scott McEathron, *Head of the Center for Graduate Initiatives and Engagement*, The University of Kansas Libraries; Julie Sweetkind-Singer, *Head Librarian, Assistant Director of Geospatial, Cartographic and Scientific Data and Services*, Stanford University Libraries and Academic Information Resources; Angela Lee, *Esri Libraries and Museum Industry Manager*; and Adrienne W. Cadle, *Senior Psychometrician* of *Professional Testing Inc*. Hallie Pritchett, *Head, Map and Government Information Library and Map and Federal Regional Depository Librarian* at the University of Georgia, was also instrumental in this project by contributing real-world examples concerning the use and management of map collections. Policy and metadata portions of the manuscript were greatly strengthened by feedback from subject matter experts Butch Lazorchak, *Information Technology Project Manager*, Library of Congress, and Matthew Mayernik, *Project Scientist and Research Data Services Specialist*, National Center for Atmospheric Research (NCAR).

This book provides a more expansive view of geographic information (GI) across the information professions, but the authors would like to recognize that many of the topics have their roots in the practice of map librarianship and the theoretical foundations of information science and library science. Therefore, a special thanks is extended to Mary Larsgaard, author of three editions of the book *Map Librarianship*. Not only is this work foundational, it continues to influence the management of print cartographic resources, including selection and acquisition; classification and cataloguing; storage, care, and repair; reference; and instruction. We also gratefully acknowledge the important and influential research published throughout the *Journal of Map and Geography Libraries*, coedited by Paige G. Andrew (The Pennsylvania State University) and Kathy Weimer (Rice University).

A special note of appreciation is extended to Marie Fazio and her team at Drexel University. Without Marie, none of this would have been possible. We would also like to thank the many graduate students and research support staff for making the GIL project a great success—Fangwu Wei (University); Sonya Prasertong (University of Kentucky); and Suzie Allard, Elizabeth Ellis, Drew Huitt, Theresa Parrish, Louisa Trott-Reeves, Bobbie Suttles, and Carol Tenopir (University of Tennessee).

Contents

List of Abbreviations

3DEP	3D Elevation Program
AAT	Art & Architecture Thesaurus
ACCR	Anglo-American Cataloging Rules
ACMLA	Association of Canadian Map Libraries and Archives
ACRL	Association of College and Research Libraries
ACS	American Community Survey
AGOL	ArcGIS Online
AJAX	Asynchronous JavaScript
ALA	American Library Association
AMS	Army Mapping Service
API	Application program interface
ARL	Association of Research Libraries
ASCII	American Standard Code for Information Interchange
ASIS&T	Association for Information Science and Technology
BLM	Bureau of Land Management
CAFF	Conservation of Arctic Flora and Fauna
CC	Core competencies
CGIS	Canada Geographic Information System
CIA	Central Intelligence Agency
CSDGM	Content Standard for Digital Geospatial Metadata
DC	Dublin Core
DCC	Digital Curation Centre
DCMI	Dublin Core Metadata Initiative
DDC	Dewey Decimal Classification
DEM	Digital elevation model
DIME	Dual Independent Map Encoding
DOM	Dissolved organic matter
DOQ	Digital orthophoto quadrangle
DwC	Darwin Core
EO	Executive order
EPA	Environmental Protection Agency

ERIC	Education Resources Information Center
EROS	Earth Resources Observation and Science Center
Esri	Environmental Systems Research Institute
FAM	Fire and Aviation Management
FCC	Federal Communications Commission
FDA	Food and Drug Administration
FDLP	Federal Depository Library Program
FEMA	Federal Emergency Management Agency
FGDC	Federal Geographic Data Committee
FOS	Free and open source
FRBR	Functional Requirements for Bibliographic Records
FTTC	Fiber-to-the-curb
GeoCONOPS	Geospatial Concept of Operations
GeoMAPP	Geospatial Multistate Archive and Preservation Partnership
Geoweb	Geospatial Web
GI	Geographic information
GIL	Geographic information librarianship/librarian
GIS	Geographic information system
GISTBoK	Geographic Information Science and Technology Body of Knowledge
GML	Geography Markup Language
GNIS	Geographic Names Information System
GOS	Geospatial One-Stop
GPO	Government Printing Office
GPS	Global Positioning System
GRASS	Geographic Resources Analysis Support System
GSDI	Global Spatial Data Infrastructure
GTCM	Geospatial Technology Competency Model
HCI	Human-computer interaction
HISB	Human information seeking behavior
ICPSR	Inter-university Consortium for Political and Social Research
IFLA	International Federation of Library Associations and Institutions
ILL	Interlibrary loan
IMLS	Institution of Museum and Library Services
IPUMS	Integrated Public Use Microdata Series
IR	Institutional repository
IRB	Institutional review board
JMGL	*Journal of Map & Geography Libraries*
KML	Keyhole Markup Language
LBS	Location-based services
LCGSA	Laboratory for Computer Graphics and Spatial Analysis
LCI	Land Cover Institute
LCSH	Library of Congress Subject Headings
LiDAR	Light Detection and Ranging
LIS	Library and information science

MAGERT	Map and Geography Round Table
MAGIRT	Map and Geospatial Information Round Table
MAUP	Modifiable areal unit problem
MCM	Map communication model
MeSH	Medical Subject Headings
MLIS	Master of library and information science
MSC	Mapping Science Committee
NACIS	North American Cartographic Information Society
NAD	North American Datum
NALCMS	North American Land Change Monitoring System
NASA	National Aeronautics and Space Administration
NBM	National Broadband Map
NBP	National Broadband Plan
NFIP	National Flood Insurance Program
NGA	National Geospatial-Intelligence Agency
NGDA	National Geospatial Data Asset
NHGIS	National Historical Geographic Information System
NIMA	National Imagery and Mapping Agency
NISO	National Information Standards Organization
NOAA	National Oceanic and Atmosphere Administration
NRC	National Research Council
NSDI	National Spatial Data Infrastructure
NTIA	National Communications and Information Administration
NWS	National Weather Service
OECD	Organisation for Economic Co-operation and Development
OGC	Open Geospatial Consortium
OMB	Office of Management and Budget
OPAC	Online public access catalog
OSGeo	Open Source Geospatial Foundation
PER	Professional emergency responders
PLSS	Public Land Survey System
POLENET	Polar Earth Observing Network
RDA	Resource Description and Access
RDF	Resource Description Framework
RF	Representative fraction
ROI	Return on investment
SCI	Science Citation Index
SDI	Spatial data infrastructure
SF	Scale factor
SPCS	State Plane Coordinate System
SPNHC	The Society for the Preservation of Natural History Collections
SVG	Scalable Vector Graphics
SYMAP	Synagraphic Mapping System
TIGER	Topologically Integrated Geographic Encoding and Referencing

TNMCorps	The National Map Corps
UCGIS	University Consortium for Geographic Information Science
UNESCO	United Nations Educational, Scientific, and Cultural Organization
UNGEGN	United Nations Group of Experts on Geographical Names
URISA	Urban and Regional Information Systems Association
USBGN	U.S. Board on Geographic Names
USDA	United States Department of Agriculture
USGS	United States Geological Survey
UTM	Universal Transverse Mercator
UXD	User experience design
VGI	Volunteered geographic information
W3C	World Wide Web Consortium
WAML	Western Association of Map Libraries
WEMI	Work, Expression, Manifestation, Item
WFS	Web Feature Service
WGS	World Geodetic System
XML	Extensible Markup Language
ZIP	Zone Improvement Plan

Chapter 1
Introduction

Abstract Geographic information (GI) generally consists of facts, data and/or evidence pertaining to events, activities and things located on (or near) the surface of the earth. The process by which humans organize, access and use this information is fundamental to science, industry, and our everyday lives. Whether this involves the use of GI to enhance facility location decisions, efficiently distribute vaccines to vulnerable populations, or to quickly navigate to the local grocery store, GI is essential. The field of Information Science provides a multidisciplinary lens to systematically study GI, its special properties and how it is accessed, processed, and used. The purpose of this chapter is to provide a brief overview of key issues surrounding GI and introduce a basic framework for understanding the input, storage, management, discovery, accessibility, and usability of GI. This chapter also emphasizes the importance of using Information Science approaches and associated theory for deepening our understanding of geographic knowledge, information, data, and related tools and provides a roadmap for the organization of this book.

1.1 Introduction

By definition, geographic information consists of facts, data and/or evidence pertaining to events, activities and things located on (or near) the surface of the earth. Geographic information (GI) is ubiquitous, persistent, and growing rapidly. GI is embedded within the daily weather forecast, it permeates many activities on social media platforms (e.g., Facebook, Twitter, etc.) and is required for any human activity that involves travel to/from origins and destinations. Increasingly, these types of spatial interactions are facilitated by global positioning systems (GPS) and navigation software, which also rely upon geographic information for processing. GI is complex. It is multidimensional, laden with context, subject to interpretation and biases, and frequently uncertain. As a result, it can be difficult to organize, manage, curate, redistribute, and analyze GI without a suite of effective strategies for dealing with its complexities.

One basic strategy used for dealing with GI in a digital environment is abstraction. In short, the abstraction process helps simplify complex information for representation, transformation, analysis, and visualization. There are two basic data models for handling the abstraction of GI. First, the *object view* considers GI as a

W. Bishop, T.H. Grubesic, *Geographic Information*, Springer Geography,
DOI 10.1007/978-3-319-22789-4_1

series of discrete entities located in geographic space. Compositionally, this includes information pertaining to the locations and shapes of geographic objects (e.g., points, lines and areas) and entities (e.g., combinations of points, lines and areas) and the relationship(s) between them (Longley, Goodchild, Maguire, & Rhind, 2015; O'Sullivan & Unwin, 2010). The compositional facets are usually stored as (1) coordinates (e.g., latitude and longitude), that function as absolute measures of geographic location, and (2) topology, which is the way in which geographic features are interrelated or arranged. Second, the *field view* considers GI to consist of properties that continuously vary across space. For example, elevation above sea level is a continuous geographic variable within the field view. Binary classifications (e.g., 0,1) for raster-based field representations are also possible, where cells have a value of 1 when certain conditions are present (e.g., forest) and 0 otherwise (O'Sullivan & Unwin, 2010).

It is also important to note that although GI is primarily focused on terrestrial events, objects and fields, *spatial* data can be connected to other spaces, not just the earth's surface (Longley et al., 2015). Thus, many of the challenges associated with GI also relate to spatial or geospatial data. This includes astronomical phenomena, allied data that were remotely sensed (Borgman, 2015), or pico-scale data such as the spatial distribution of bacterial colonies and their habitats in the human body (Costello et al., 2009).

Again, the challenges associated in working with GI are numerous, including multidimensionality, voluminosity, projecting data onto a flat surface, varied information types, complex data formats, dynamism, place-name changes and toponymic homonyms, and domain-specific metadata for appraisal and subsequent use (Larsgaard, 2005; Maguire & Longley, 2005; Morris, 2013). To make matters more difficult, much of GI is proprietary.

Given the complex nature of GI, its wide use and relative importance to so many domains in the public and private sectors, the purpose of this opening chapter is to briefly detail the many challenges associated with interacting with GI. This includes an overview of a basic framework for understanding the input, storage, management, discovery, accessibility, and usability of GI. We also provide short introductions to relevant information policies concerning GI, highlight industry connections to GI, and briefly highlight the dynamic nature of GI education. We conclude this chapter with a roadmap for the organization of this book.

1.2 Geographic Information and the Data Deluge

One significant challenge associated with GI is that much of the core data required for decision-making is difficult and/or costly to collect (e.g., population census, satellite imagery, and so forth). As a result, policymakers and researchers have actively sought to craft procedures for sharing data. Not only does this include the development of processes and standards to simultaneously advance science through the use of data, but also to remove duplicative efforts in data capture, storage, and distribution efforts. Historically, the role of GI collection fell to governments with the authority and resources to collect large amounts of geospatial data. These data were often critical to national security, informing defense efforts, or important for local government agencies to provide support services to their local population. As a result, governments largely dictated the collection, maintenance, and availability of these data (Rhind, 1999). The "data deluge" described by Hey and Trefethen (2003) has become accepted as a fundamental characteristic of science today, as the scientific data corpus continues to increase at a rate of around 30% per year (Pryor, Jones, & Whyte, 2014). However, with the commercialization of the Internet, the wide availability of global positioning receivers, and the growth of the Geoweb (Haklay, Singleton, & Parker, 2008), this data deluge is more expansive (and more spatial) than ever. In short, the ability to amass and share GI has become more democratized. Any individual operating a mobile device capable of using location-based services (LBS) or equipped with a GPS receiver or conduct their own remote sensing with drones can collect and contribute user-generated data, widely labeled as *volunteered geographic information* (VGI) (Goodchild, 2007).

What makes the emergence of the Geoweb, LBS, and related applications so important is that the creation, manipulation, and revision of GI can be done more efficiently and inexpensively than traditional (e.g., government) approaches. More importantly, all of this can be done in real-time. Of course, VGI, and the Geoweb also have different standards for access, authority, and quality controls, but this does not encumber the generation of data (Elwood, Goodchild, & Sui, 2013). The exponential rate of increase in data from user-generated content and associated research output, as well as the continued production of open government data, presents new challenges for those charged with storing, organizing, and making GI discoverable. In fact, many significant challenges related to this information exist, including data curation, data provenance, portability, and transferability. In essence, GI must retain characteristics that allow for its use beyond its original creators and associated purposes, but achieving these goals is no easy task.

In a nutshell, these are the core problems to be explored in this book. We are concerned with how knowledge organization, information retrieval, and human information seeking behavior impacts GI organization, access, and use. Thus, unlike other books on GI or its analysis, the purpose of this work is to provide an introduction to the management, curation, and preservation of GI for future access, use, and reuse. The information management decisions made by creators at each step of the data lifecycle impact how users will retrieve and use GI beyond its original purposes.

Further, it is also important to note that the field of Information Science (Borko, 1968) transcends all scales and characteristics of geospatial data. As a result, Information Science provides a multidisciplinary lens to systematically study GI from information systems, sciences, and studies perspectives. Because research related to the input, storage, management, discovery, accessibility, and usability of GI undergirds work in so many disciplines (e.g. Geography, Criminology, Epidemiology, Urban Planning, Physics, and so forth), information structures, information seeking behaviors, information use, and associated educational efforts have dramatic impacts on how place-based decision-making occurs. Further, all of these factors impact the development of meaningful public policy and the process of scientific discovery, more generally. Information Science provides a wide, yet focused lens for pursuing these important questions across disciplines, sectors, and communities, without the need for each to reinvent their own approaches.

1.2.1 Geographic Information Policy

The shift between analog and digital geospatial resources was greatly accelerated in the U.S. by the National Research Council's (1993) request for the development of a spatial data infrastructure (SDI), as well as President Clinton's Executive Order No.12906 (1994), forming the National Spatial Data Infrastructure (NSDI). Broadly, an SDI requires coordination in the creation, collection, dissemination, and storage of spatial data between stakeholders in a spatial data community (Williamson, Rajabifard, & Binns, 2006). These national efforts for establishing SDIs roughly coincided with the National Research Council's U.S. Mapping Science Committee (MSC) report entitled *Toward a Coordinated Infrastructure for the Nation* that suggested the creation of an infrastructure for the nation's geospatial data, including the collection of the "materials, technology, and people necessary to acquire, process, and distribute" that geospatial data (National Research Council [NRC], 1993, p. 16) was necessary. To their credit, the MSC recognized the existence of an unregulated and incoherent *ad hoc* geospatial data infrastructure, but charged that the creation of a coordinated national infrastructure designed to facilitate the use of GI among stakeholders would help assist in solving important place-based problems. In particular, Executive Order No.12906 (1994) stressed the importance of *coordinated* development, use, sharing, and dissemination of digital GI on a national basis in order "to promote economic development, improve our stewardship of natural resources, and protect the environment" (p. 17671). This geographic information policy also spawned the creation of metadata standards from the Federal Geographic Data Committee (FGDC), prompting the need for geolibraries to facilitate sharing of geographic information (Goodchild, 2009). The cyberinfrastructure of the NSDI vision expands and echoes this in Executive Order No. 13,642 (2013) stating that "openness in government strengthens our democracy, promotes the delivery of efficient and effective services to the public, and contributes to economic growth" (p. 244).

The reason readers need to be aware of these U.S. federal efforts is three-fold. First, to make all GI open, the data must be collected, documented, organized, managed, and preserved. In this context, the workforce of libraries, archives, museums, and data centers often serve as *information intermediaries*—brokering products with little input on the decisions concerning format, metadata, and delivery of this information. Second, because government GI and other data on the Geoweb are made available online, both users and information agencies do not necessarily retain local copies of geospatial data and e-government and many online materials are not permanent (Bishop, Grubesic, & Prasertong, 2013). The sheer quantity of U.S. federal GI via the NSDI provides standardization, a framework, and a clearinghouse of data needs information professionals to improve GI organization, access, and use, and these same practices may be applied to all other GI types. Third, information generated and consumed via the Geoweb provides almost limitless opportunities for contributors to create, revise, and remove GI without the authority or training of an information professional. These data provenance issues drastically impact the findability, usability (e.g., metadata), and continuity of these data. Once again, these gaps in policy, procedure, and process require mitigation.

1.2.2 *The Data Lifecycle*

Similar to any other public or private asset, data have a lifecycle and require informed management strategies. In this context, an important and widely used model is the Digital Curation Centre's (DCC) Curation Lifecycle Model. The DCC Lifecycle Model identifies the process by which data is either disposed of, or selected for reuse and long-term preservation (Higgins, 2008). Throughout this book, we borrow liberally from the DCC model and its associated terminology to better frame the discussion of managing GI, making sure to cross-reference all steps within the DCC lifecycle.

In the DCC model, all data begins at the conceptualization stage. Conceptualization of GI is done every day, by everyone on the planet when attempting to decipher and make sense of the spaces and places around them. This type of spatial reasoning is conducted when walking around one's house or apartment, commuting on a bus, or navigating to a local market to purchase groceries. However, there is also a subset of the population that creates and analyzes GI for a living. This includes geographers, cartographers, surveyors, or those working in Natural Resources Management, Urban Planning, Environmental Science, Biology, Ecology, Hydrology, Agriculture, Forestry, Engineering, Geology, the medical and health fields, Transportation, Atmospheric Science, the Military, Marketing Research, Logistics, Education, Archaeology, Landscape Architecture, Criminology, and scholars in the emergent Geohumanities (Ayers, 2010). Beyond the conceptualization stage, the primary platform in which professionals analyze, manipulate, transform, and visualize GI is a geographic information system (GIS) or related statistical computing platforms such as R.

Aside from the fundamental tasks associated with the conceptualization of data, most GIS users engage in at least some of the following GI-related activities: (1) creating or receiving data; (2) appraising and selecting data; (3) ingesting; (4) preserving; (5) storing; (6) accessing, using, and reusing data; and (7) transforming data. Unfortunately, the knowledge, skills, and abilities required to conduct many of these tasks are not part of the typical GIS user's toolkit. This is a critical gap for many reasons. Consider, for example, that in 2011, the National Science Foundation (NSF) began requiring data management plans, and other agencies such as the National Institutes of Health (NIH) and the National Aeronautics and Space Administration (NASA) quickly followed suit. Thus, although formal policies require compliance from funded projects to ensure data sharing and preserving standards, as well as data lifecycle management, many GI scientists and other GIS users lack the human and professional resources to develop meaningful strategies to accomplish these tasks. These resource gaps require mitigation.

It is also important to note that not all GI requires preservation. For example, privacy concerns stemming from surveillance data (Lyon, 2003) should engender a response to destroy, not preserve, personally identifiable GI, but this is not always the case (Crampton, 2008; Dodge, Batty, & Kitchin, 2004). Thus, part of the data lifecycle management for sensitive data should require strategies for disposal. Similarly, there is a large body of GI that is not intended to be open nor redistributed because of security concerns or allied privacy issues. As a result, access and use issues vary by dataset and any good data management plan or strategy should reflect these constraints accordingly. All the activities in the data lifecycle occur with any GI, but for some sets of data information policy exists to guide these activities more prescriptively than others.

1.2.3 The GI Industry and Education

The jobs related to managing this universe of GI are still emerging, but many sources agree that employment in these domains is growing. The Occupational Outlook Handbook states that employment in fields which use GIS (e.g., geographic information specialist) are expected to increase at a rate faster than the average rate of growth for all occupations between 2014 and 2024 (i.e., increase 29%) (U. S. Department of Labor. Bureau of Labor Statistics, 2016). This translates to an expected increase of at least 100,000 or more job openings in the U.S. over the next decade.

The Boston Consulting Group (2012) estimates that the U.S. geospatial industry generated $73 billion in revenue in 2013, with 500,000 high-wage jobs. Oxera (2013) puts the global revenue number at up to $270 billion per year. These Google-funded studies define the geospatial industry beyond just those professionals creating and using GI to include the entire value chain of geoservices from geospatial data collection to other related location-based services (e.g., satellites, location-based search, and so forth). In order to support this rapid growth, a workforce is needed that understands geospatial technology and the associated tasks related to the GI input, storage, management, discovery, accessibility, and usability.

In recent years, significant efforts have been made to design GIS and technology (GIS&T) curricula that aligns with workforce needs, especially to meet demand for professionals that create and analyze GI (DiBiase, 2007; Estaville, 2010; Wikle, 2010). A body of existing GIS coursework exists within many colleges and universities in the U.S. and abroad, including departments of Geography, Urban Planning, Public Policy, Natural Resource Management, Environmental Science, Archaeology, Anthropology, Economics, Political Science, History, Humanities, Education, Business, and homeland security programs, as well as many technical and vocational institutions, and about 450 community colleges (GeoTech Center, 2010; Unwin, 2011). The GIS Certificate Institute (GISCI) facilitates a program for Certified GIS Professionals (GISP) to standardize the knowledge and experience of people working in these fields. Although training related to GIS, remote sensing, and GPS technologies remains central to GI use, education related specifically to the information management issues of GI is as absolutely critical to the efficient and effective integration of GI throughout multiple domains. The need for further curriculum development based on real-world practice could be synergistic with other programs such as those found in the iSchools (http://ischools.org/).

In sum, careers in GI have the potential to attract a wide range of individuals across numerous fields, but an education in data science, library science, archiving, museum studies and digital curation can provide professionals with a unique skill set that will help steer the future of GI organization, access, and use. By maximizing the potential of GI openness through improvements in the current discoverability, accessibility, and usability of tools and data, the vision of a unified, global, SDI can be realized.

1.3 Book Organization

Given the issues outlined above, the principle aim of this book is to provide readers with an introduction to the management and preservation of GI created by an array of organizations and a wide range of geospatial tools. At its core, this book is intended to serve as a resource for educators and students who work with GI by addressing the knowledge, skills, and abilities required to efficiently create, access, organize, curate, and redistribute GI. Throughout, we address several core themes, including: (1) the roles of librarians, archivists, data scientists, and other information professionals in the creation of GI representations for its organization, access, and use; (2) we draw upon the rich history of GI creation and storage, including key geographic and cartographic elements; (3) we cover the beneficial background related to geographic information policy and metadata creation; (4) we endeavor to provide a detailed overview of data discovery, fitness for use, human information seeking behavior, and the data lifecycle processes; and (5) we also provide an educational roadmap in the milieu. With these goals in mind, the ten chapters of this book are organized into three different sections:

- Part I—Organization
- Part II—Access and Use
- Part III—Education

In the next chapter, we provide an overview of the core components of GI, a review of map functions and an abbreviated history of GIS. This chapter is written for readers needing a quick refresher on key concepts in geography and the spatial sciences, emphasizing the importance of maps as repositories for GI and the primary tool used for mapmaking today, GIS. Although this information is available from other sources, it is widely scattered and we believe that many people who work with GI on a daily basis that do not have formal training in geography would benefit from a brief review.

Chapter 3 delves deeper into the key elements of most GI, including issues of geographic scale, projections, and coordinate systems—all of which coalesce to form key elements of geospatial metadata.

Chapter 4 explores key elements of geographic information policy, their associated motivations and implications. However, because GI and its use are so widespread, SDI programs are explored broadly in addition to U.S. policy.

Part I of the book concludes with Chap. 5, which provides a review of metadata, its value, and purposes as well as a review of other knowledge organization concepts and a few examples of geospatial metadata schemas, profiles, and standards.

Part II of this book changes course somewhat, providing an overview of the user-side of Information Science, focusing on issues of information retrieval, accessibility, usability, and their enabling technologies. Specifically, Chap. 6 provides a broad overview of the platforms and data that make the Geoweb possible. It also explores the issues stemming from authority on the production and consumption of VGI and usability, functionality, and accessibility implications for equitable GI use. Chapter 7 presents a substantial list of authoritative resources that are needed to obtain imagery, maps, and other geospatial data. The concept of *fitness for use* is included to frame appropriate selection of GI for use in different contexts. Chapter 8 focuses on human information seeking behavior related to the information needs, uses, and software choices that relate to GI access and use. Part II of the book concludes with Chap. 9, which explores the data lifecycle. A section on analog collection development and maintenance is included as many information professionals working with digital objects also archive print cartographic resources.

Finally, Part III of this book outlines both the current environment (opportunities and threats) and a path forward in educating for the different occupations in GI organization, access, and use. Not surprisingly, there are significant educational opportunities in these emerging markets and Chap. 10 presents a multidisciplinary approach to developing a curriculum for the geoservices workforce by coordinating with existing curricular scaffolds from K-12 to higher education.

References

Ayers, E. (2010). Turning toward place, space and time. In D. Bodenhamer, J. Corrigan, & T. Harris (Eds.), *The spatial humanities: GIS and the future of humanities scholarship* (pp. 1–8). Bloomington, IN: Indiana University Press.

Bishop, B. W., Grubesic, T. H., & Prasertong, S. (2013). Digital curation and the GeoWeb: An emerging role for geographic information librarians. *Journal of Map and Geography Libraries, 9*(3), 296–312. doi:10.1080/15420353.2013.817367.

Borgman, C. L. (2015). *Big data, little data, no data: Scholarship in the networked world.* Cambridge, MA: MIT Press.

Borko, H. (1968). Information science: what is it? *American Documentation, 19*(1), 3–5.

Boston Consulting Group. (2012). *Putting the U.S. geospatial industry on the map.* Retrieved October 30, 2015, from http://www.ncge.org/files/documents/US-FullReport.pdf.

Costello, E. K., Lauber, C. L., Hamady, M., Fierer, N., Gordon, J. I., & Knight, R. (2009). Bacterial community variation in human body habitats across space and time. *Science, 326*(5960), 1694–1697.

Crampton, J. W. (2008). The role of geosurveillance and security in the politics of fear. In D. Z. Sui (Ed.), *Geospatial technologies and homeland security* (pp. 283–300). Dordrecht, The Netherlands: Springer.

DiBiase, D. W. (2007). Is GIS a wampeter? *Transaction in GIS, 11*(1), 1–8. doi:10.1111/j.1467-9671.2007.01029.x.

Dodge, M., Batty, M., & Kitchin, R. (2004). *No longer lost in the crowd: Prospects of continuous geosurveillance. Association of American Geographers Annual Conference.* Philadelphia, PA.

Elwood, S., Goodchild, M. F., & Sui, D. (2013). Prospects for VGI research and the emerging fourth paradigm. In D. Sui, S. Elwood, & M. F. Goodchild (Eds.), *Crowdsourcing geographic knowledge: Volunteered geographic information (VGI) in theory and practice* (pp. 361–375). New York: Springer. doi:10.1007/978-94-007-4587-2_20.

Estaville, L. E. (2010). Geospatial workforce trends in the United States. *International Journal of Applied Geospatial Research, 1*(1), 57–66.

Exec. Order No. 12906, 3 C.F.R. 17,671–17,674 (1994).

Exec. Order No. 13642, 3 C.F.R. 244–246 (2013).

GeoTech Center. (2010). *National Geospatial Technology Center of Excellence*. Retrieved from http://www.geotechcenter.org

Goodchild, M. F. (2007). Citizens as sensors: the world of volunteered geography. *GeoJournal, 69*(4), 211–221.

Goodchild, M. F. (2009). Geographic information systems and science: Today and tomorrow. *Annals of GIS, 15*(1), 3–9.

Haklay, M., Singleton, A., & Parker, C. (2008). Web mapping 2.0: The neogeography of the GeoWeb. *Geography Compass, 2*(6), 2011–2039. doi:10.1111/j.1749-8198.2008.00167.x.

Hey, T., & Trefethen, A. (2003). The data deluge: An e-Science perspective. In F. Berman, G. Fox, & T. Hey (Eds.), *Grid computing: Making the global infrastructure a reality* (pp. 809–824). Chichester, England: Wiley.

Higgins, S. (2008). The DCC curation lifecycle model. *International Journal of Digital Curation, 3*(1), 134–140 Retrieved from http://www.ijdc.net/index.php/ijdc/article/viewFile/69/48.

Larsgaard, M. L. (2005). Metaloging of digital geospatial data. *Cartographic Journal, 42*(3), 231–237.

Longley, P. A., Goodchild, M. F., Maguire, D. J., & Rhind, D. W. (2015). *Geographic information science and systems*. Hoboken, NJ: Wiley.

Lyon, D. (2003). Surveillance as social sorting: Computer codes and mobile bodies. In D. Lyon (Ed.), *Surveillance as social sorting: Privacy, risk and digital discrimination* (pp. 13–30). London: Routledge.

Maguire, D. J., & Longley, P. A. (2005). The emergence of geoportals and their role in spatial data infrastructures. *Computers, Environment and Urban Systems, 29*, 3–14.

Morris, S. (2013). *Issues in the appraisal and selection of geospatial data: An NDSA Report*. NSDA Content Working Group. Retrieved from http://hdl.loc.gov/loc.gdc/lcpub.2013655112.1.

National Research Council (1993). *Toward a coordinated spatial data infrastructure for the nation*. Washington, DC: National Academies Press.

O'Sullivan, D., & Unwin, D. J. (2010). *Geographic information analysis* (2nd ed.). Hoboken, NJ: Wiley.

Oxera Consulting, Ltd. (2013). *What is the economic impact of geo services?* Retrieved from http://www.oxera.com/Oxera/media/Oxera/downloads/reports/What-is-the-economic-impact--of-Geo-services_1.pdf?ext=.pdf.

Pryor, G., Jones, S., & Whyte, A. (2014). *Delivering research data management services: Fundamentals of good practice*. London: Facet Publishing.

Rhind, D. (1999). National and international geospatial data policies. In P. A. Longley, M. F. Goodchild, D. J. Maguire, & D. W. Rhind (Eds.), *Geographical information systems—Principles and technical issues* (Vol. 1, pp. 767–787). New York: Wiley.

U. S. Department of Labor. Bureau of Labor Statistics. (2016). *Occupational outlook handbook*. Retrieved from http://www.bls.gov/ooh/.

Unwin, D. (2011). *Teaching geographic information science and technology in higher education*. Hoboken, NJ: Wiley.

Wikle, T. (2010). Planning considerations for online certificates and degrees in GIS. *URISA Journal, 22*, 21–30.

Williamson, I., Rajabifard, A., & Binns, A. (2006). Challenges and issues for SDI development. *International Journal of Spatial Data Infrastructures Research, 1*, 24–35.

Chapter 2
Geographic Information, Maps, and GIS

Abstract The purpose of this chapter is to provide readers with an overview of the core components of geographic information (GI), a review of map functions, and an abbreviated history of geographic information systems (GIS). The origins of the map, map functions, and a discussion of place and space will provide a foundational understanding of important concepts for information professionals working with GI. This includes providing readers with a basic awareness of how maps serve as repositories for GI and its communication. Finally, a brief timeline of modern GIS is detailed, providing readers with a basic perspective on the emergence of these important tools and their role in shaping geographic information production and consumption.

2.1 Introduction

Attempting to discuss geographic information (GI) without referring to maps is analogous to discussing bibliographic information without referring to books. Maps are important, useful, and they function as repositories for vast amounts of GI. Thus, when writing a chapter that deals with the basics of GI and its storage, the benefits of discussing maps, at least in some detail, are obvious.

Based on an array of evidence from archeology, semiotics, linguistics, and human cognitive development, scientists believe that maps were a precursor to mathematics and written language, but were developed after music and dance (Martin, 2005). This means that maps are likely one of the oldest types of information constructed. Maps often provide people with a relatively straightforward progression from reality to abstract representation, minimizing information loss, and confusion during the transfer process. Archeologists found a simple example from nearly 14,000 years ago in a cave in northern Spain. The tablet depicts the entrance to a cave, with a series of wavy lines to represent a river, and other animals and features are drawn in relation to the cave (Utrilla, Mazo, Sopena, Martínez-Bea, & Domingo, 2009). This simple map would not appear much different from one produced today by a person using pen and paper with no formal training in Geography or Cartography.

Similar to any means of communication, maps have evolved over time. From 234 to 192 B.C.E., Eratosthenes worked as the chief librarian of the Alexandria Library in Alexandria, Egypt. Not only was Eratosthenes the first person to roughly

W. Bishop, T.H. Grubesic, *Geographic Information*, Springer Geography,
DOI 10.1007/978-3-319-22789-4_2

calculate the circumference of the earth, but he also coined the term "Geography." Although Eratosthenes' efforts in determining the circumference of the earth are covered in detail elsewhere (Dutka, 1993; Fischer, 1975), the basics are worth recounting for readers.

Eratosthenes had learned of a deep well in Syene, which is known today as Aswan. The sun shown on the bottom of the well only one day per year, June 21, which corresponded to the summer solstice. Thus, he accurately surmised that the sun was directly over the well on June 21. By measuring the angle of the sun above the horizon in Alexandria at noon on June 21, which he believed was due north of Syene, and calculating the length of a shadow cast by a vertical column of known height and representing this shadow as a vertical line, he determined the vertical angle was 82°48'. When combined with the known sun angle at the well in Syene (90°00'), Eratosthenes extended these vertical lines to the earth's center and they formed an angle of 7°12'. This meant that the arc distance between both locations, relative to the earth's circumference must be 7°12'/360°, or 1/50th of the circumference (Robinson, Morrison, Muehrcke, Kimmerling, & Guptill, 1995). The final step in his calculation was determining the distance between Alexandria and Syene, which was estimated to be about 5000 stadia (i.e., 925 km or 575 miles). At the time Eratosthenes was working on his calculations, a Ptolemaic stadia was about 185 m. Since this corresponded to 1/50th of the earth's circumference, Eratosthenes multiplied by 50 and estimated that the earth's circumference was approximately 250,000 stadia, or 46,250 km, or 28,750 miles. Eratosthenes' calculations were not exactly correct, but his basic methodology and associated process set the stage for formalizing earth measurements.

Although measuring the earth's circumference seems to be a rather trivial task today, it is important to remember that the early scientists believed the earth to be a relatively simple entity, not necessarily flat, with many placing Jerusalem as its relative center. To those working at Alexandria with the knowledge gathered there, much of the world was still unknown. Eratosthenes in his three-volume tome *Geographika*, summarized and criticized earlier works, described how he measured the circumference of the earth, and used parallels to divide his political maps into a grid. In these works lay the seed of further study related to both Physical Geography and Human Geography. We will return to the basics of georeferencing and measuring the earth in Chap. 3, but the symbiotic concepts of place and space will be addressed later in this chapter. In the meantime, it is important to reiterate how and why maps continue to function as key repositories of GI and a means for communicating this information widely.

2.2 A Constellation of Maps

As mentioned earlier in this chapter, maps are more than simple abstractions or reductions of real-world GI. Robinson et al. (1995, p. 10) suggest that a map is a "carefully designed instrument for recording, calculating, displaying, analyzing, and understanding the interrelation of things." Also, all maps, regardless of their intended function, have two things in common—representing *locations* and *attributes*. Locations correspond to positions on the earth's surface and are typically represented using *x*,*y* coordinates. Attributes include information about the characteristics or nature of a particular location (e.g., annual precipitation, number of crimes, and so forth). Although simple, these two basic elements of GI can be used to derive a suite of relational information from similarly referenced features. This may include basic distance calculations between two locations, tabulating the differences in annual snowfall levels between locations, and correlating the relationship between snowfall and proximity to one of the Great Lakes, exploring the connection between poverty and education levels across U.S. counties, or using these elements for deriving important topological information related to size, shape, contiguity, adjacency, and so forth.

2.2.1 Scale

One of the more confusing elements of a map, especially for readers that are new to formalized representations of GI, is scale. When real-world geographic features (e.g., lakes, rivers, and so forth) are reduced in size for map placement, the ratio between reality, and the reduced map dimensions is referred to as *map scale*. Small scale maps show a large geographic area on a small sheet of paper or screen. For example, Fig. 2.1 illustrates the lower 48 states at a scale of 1 in. = 426136 miles, or 1 in. = 27,000,000 inches on the map. In general, small-scale maps reduce detail for

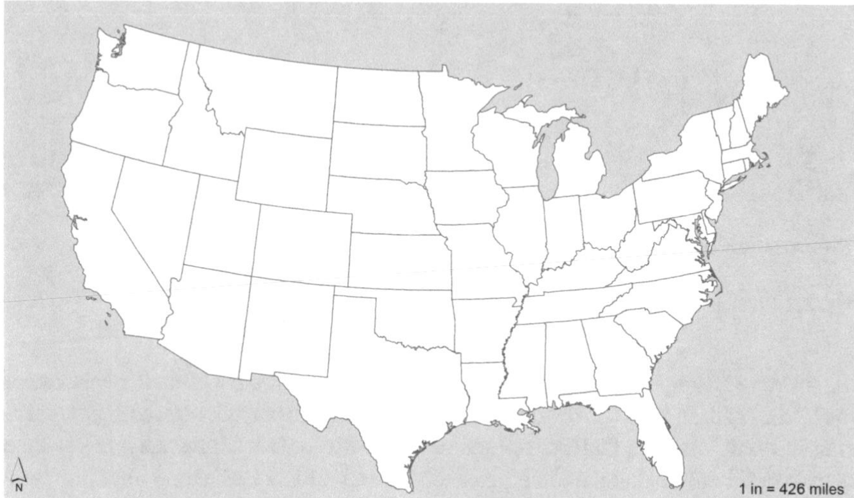

Fig. 2.1 The United States

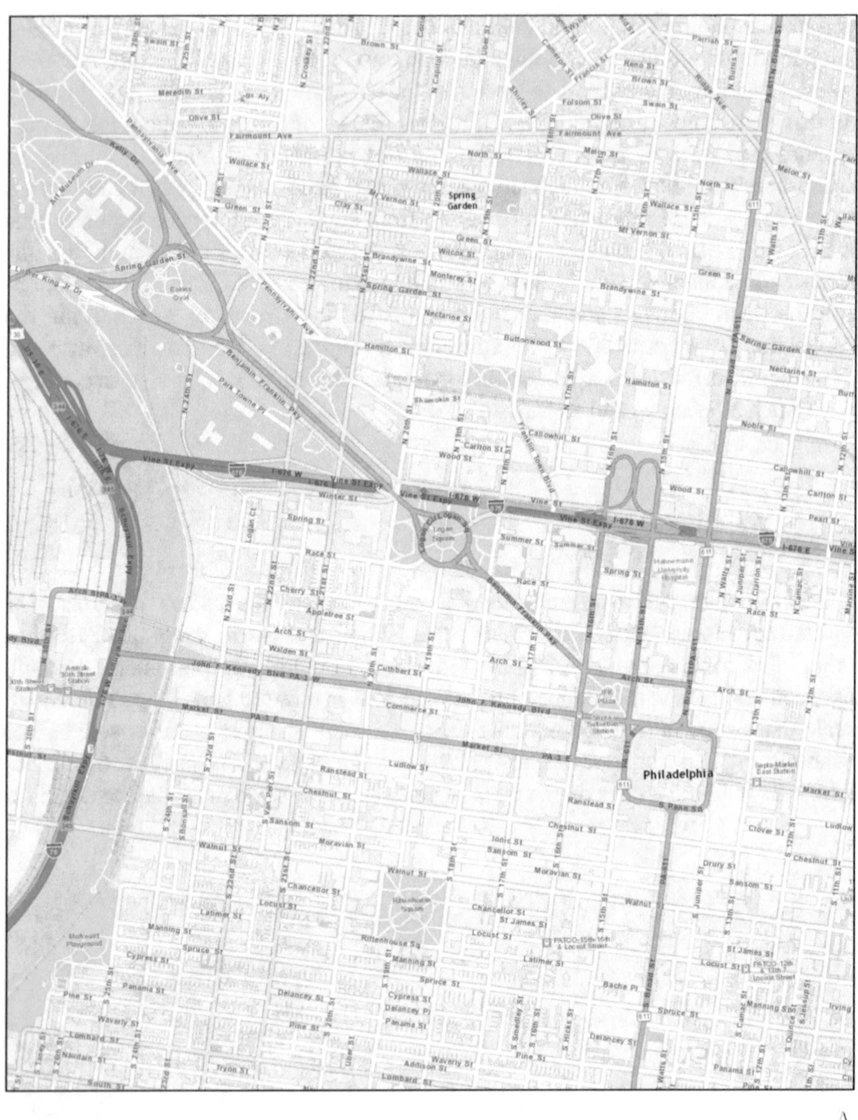

Fig. 2.2 Center City, Philadelphia

the purpose of emphasizing important features and providing a holistic overview of the GI in hand. In Fig. 2.1, basic state boundaries are displayed, but no other significant information (e.g., features or landmarks) is illustrated. Conversely, large-scale maps provide more detail and emphasize a variety of features, some important and others not. For example, Fig. 2.2 illustrates a portion of the city of Philadelphia at a scale of 1 in. = .176 miles or 1 in. = 11,168 in. on the map. A variety of important features are detailed, including the Schuylkill River, Logan Square, and City Hall, as well as many less important features, such as the building footprints of private residences and many minor streets. Lastly, where scale is concerned, it is important for readers to remember that scale refers to the relative sizes of object representations on the map, not the amount of reduction that takes place between reality and the map (Robinson et al., 1995).

2.3 Map Functions

Maps have a variety of functions, from providing general reference to exploring a particular theme or for being used as a navigational aid. Understanding the differences in map functions is critical for several reasons. First, no map, regardless of its structure or the underlying information, can hope to convey, visualize, or address every question that a reader may have about a location or its attributes. Good maps are limited in both scale and scope, focusing their communicative efforts on a limited set of important GI. Second, there is context associated with representing GI. Thus, in addition to the basic cartographic literacy required to read, interpret, or create a map, readers should also have enough situational awareness to discern which type of map should be used for appropriately conveying GI. In this section, we take a closer look at three general classes of maps.

2.3.1 Reference Maps

Reference maps, including the first known map (Utrilla et al., 2009), are structured to locate and identify important real-world features, such as water bodies, coastlines, cities, roads, or buildings. As noted by Kimmerling, Buckley, Muehrcke, and Muehrcke (2011), reference maps purposely attempt to maximize detail and minimize error so that the information portrayed in the map can be used with confidence. Moreover, all features are given equal visual prominence. For example, Fig. 2.3 displays a basic reference map for the Southwestern United States and the Interstate 10 corridor between Phoenix, Arizona and Los Angeles, California. In addition to cities, water features, state boundaries, national boundaries, state highways, and interstate highways, local geophysical relief is also displayed. A second reference map, Fig. 2.4, is a topographic map that highlights many natural and cultural features. In this case, the San Francisco Peaks, which are north of Flagstaff, Arizona, are illustrated. In addition labeling the many springs in the area, the topographic map also demarcates a ski lift (running up the western slope of Agassiz Peak), several roads, and a pipeline. Topographic maps also include contour lines for representing elevation. Clearly seen in Fig. 2.4 is the 3000 m contour (9842 ft.), and several points demarcating local peaks. For example, Humphreys Peak, the highest peak in Arizona is 3851 m (12,635 ft) tall.

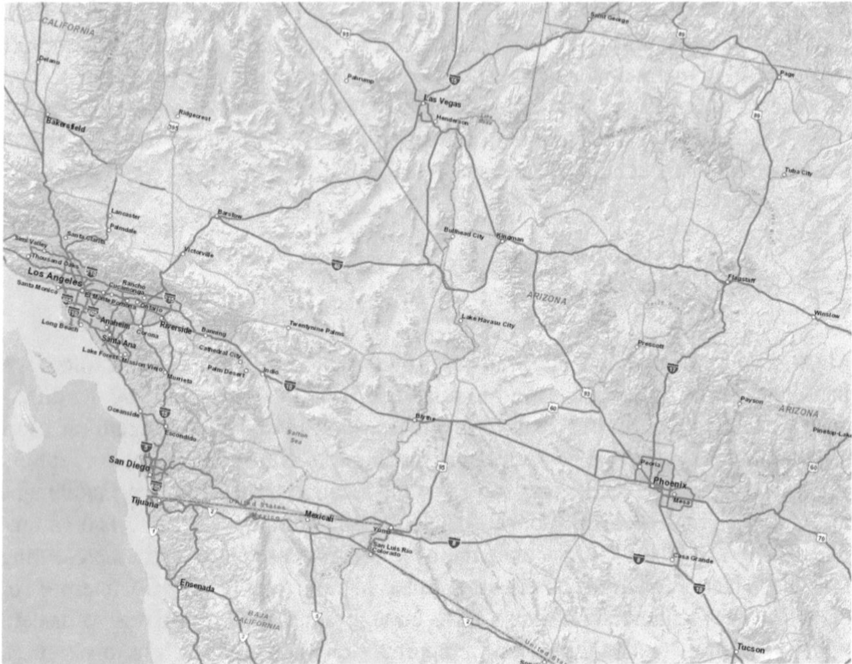

Fig. 2.3 Southwestern United States

2.3.2 Thematic Maps

Thematic maps, unlike reference maps, are structured to emphasize particular features or events. This often includes the distribution of an attribute and its relationship to other features. There are many different types of thematic maps; isopleth, choropleth, dot density, graduated symbol, proportional symbol, cartograms, and even mental maps can be considered thematic maps. Consider, for example, Fig. 2.5 which displays household median income levels for the Portland, Oregon area using Census tracts. In this map, income levels are comparable across the metropolitan area. In other words, readers can explore, compare, and draw conclusions regarding spatial distribution of income levels for the region. Similarly, Fig. 2.6 highlights average annual precipitation levels for the state of Arizona. Here, it is clear that the desert valleys in the southwestern portion of the state have extremely low precipitation values relative to the mountainous regions in the north-central portions of Arizona. It is important to note, however, that maps dealing with a single class of phenomena are not necessarily thematic in nature. For example, if a map of population density is structured simply to show the locations of urban centers, it might be more applicable to classify it as a general reference map. That said, if the emphasis is to focus attention on the spatial distribution of population density, then the map is considered thematic (Robinson et al., 1995).

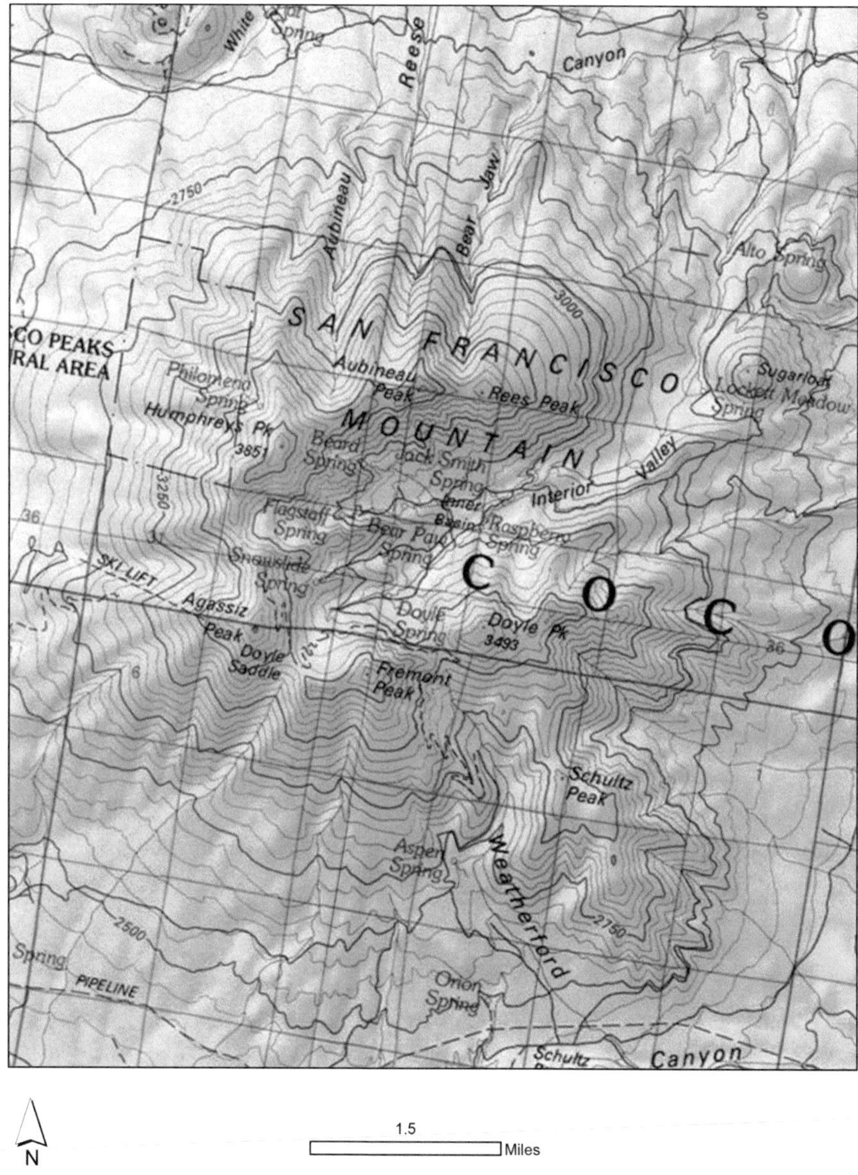

Fig. 2.4 USGS Topographic Quadrangle for the San Francisco Peaks, Arizona

2.3.3 *Charts*

A final class of maps, known as charts, are typically nautical or aeronautical in structure, designed to help users navigate over land, water, or through the air. Although most users simply look and/or examine reference and thematic maps, charts provide a foundation for actively engaging and working on the map. Routes

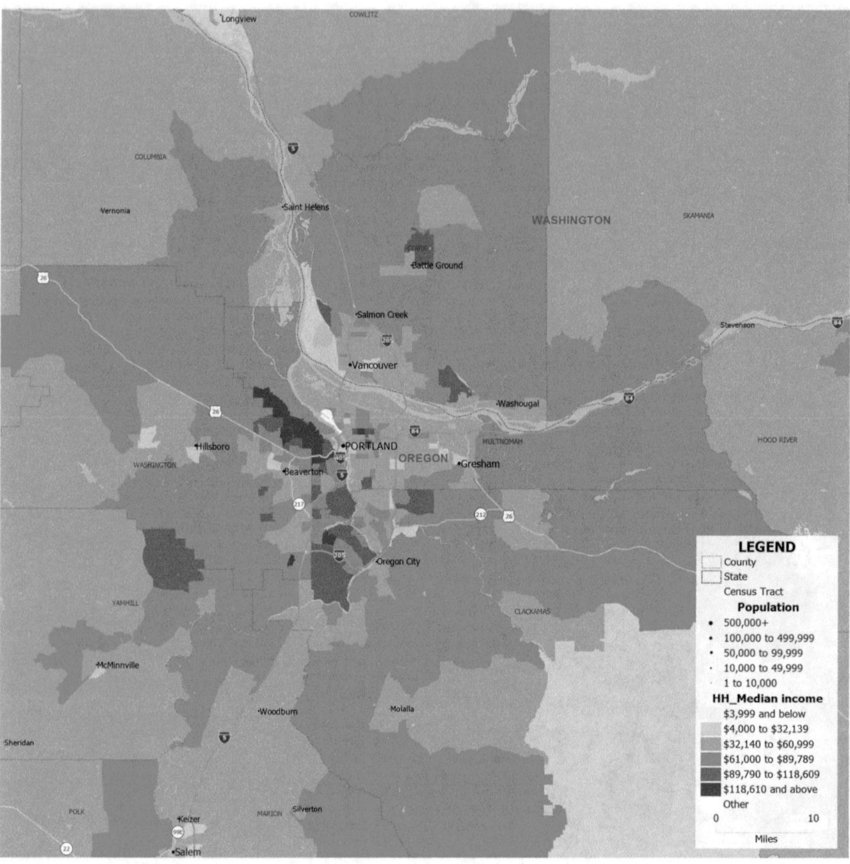

Fig. 2.5 Median Household Income, Portland, Oregon, 2010.

are planned, bearings are marked, and prominent visual features for the journey are denoted. Much like general reference maps, charts are designed to enhance visual ease of use. Figure 2.7 displays a portion of the Federal Aviation Administration aeronautical chart for the Cincinnati, Ohio region at an original scale of 1:500,000. Much of the information one would expect from an aeronautical chart is present, populated place outlines, low altitude air routes, obstructions, water bodies, and a range of other geophysical and cultural features. Figure 2.8 illustrates a portion of the nautical chart for the San Francisco Bay. Again, many of the features that one would expect are present, including shipping lanes, regulated navigational areas, navigational hazards, deep water routes. The common theme to most charts, aside from their general purpose, is keeping operators, passengers, and cargo safe from hazards. This is one of the reasons that navigational obstructions are so prominently displayed on these charts.

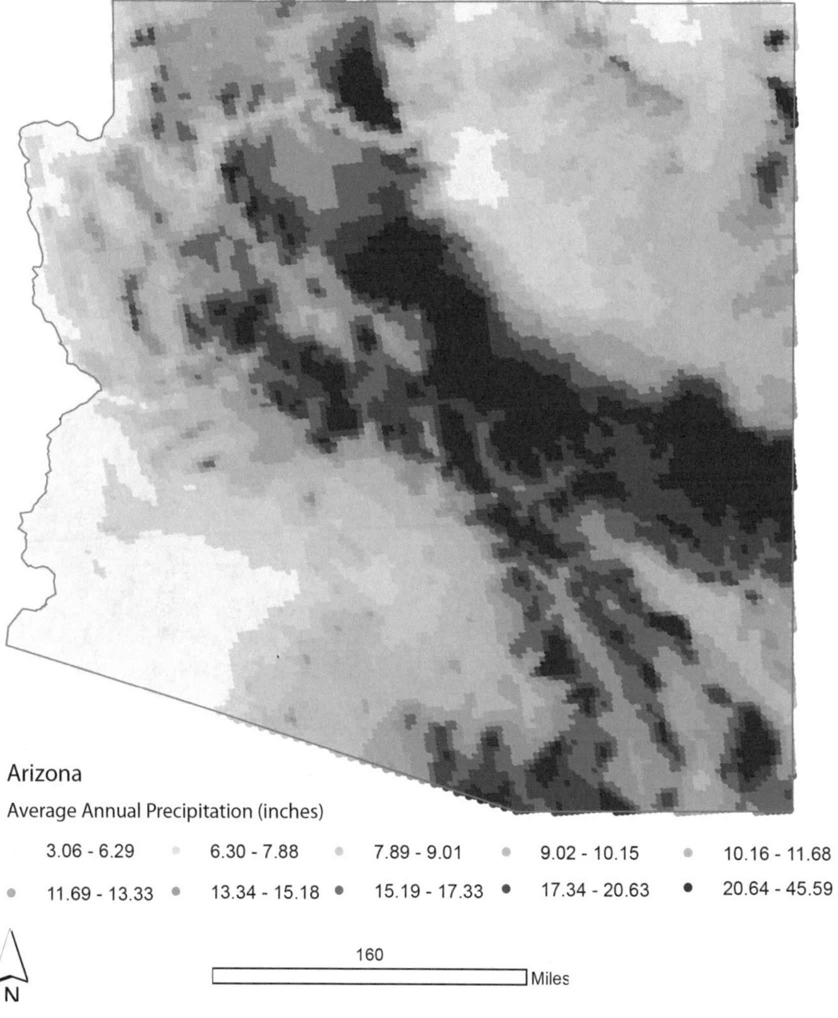

Fig. 2.6 Average annual precipitation levels for the state of Arizona

2.4 Space and Place

As detailed above, there are many different types of maps, map functions, and approaches for visualizing GI. Primary undercurrents to this process are the symbiotic concepts of *space* and *place*. By definition, space refers to the dimensions of height, depth, and width within which all things exist and move (Oxford Dictionaries, 2016). As detailed previously, geographic space corresponds to positions at or near the earth's surface, but it need not be limited to the earth. A common representation of geographic space are Cartesian coordinates, which form a grid that allows one to

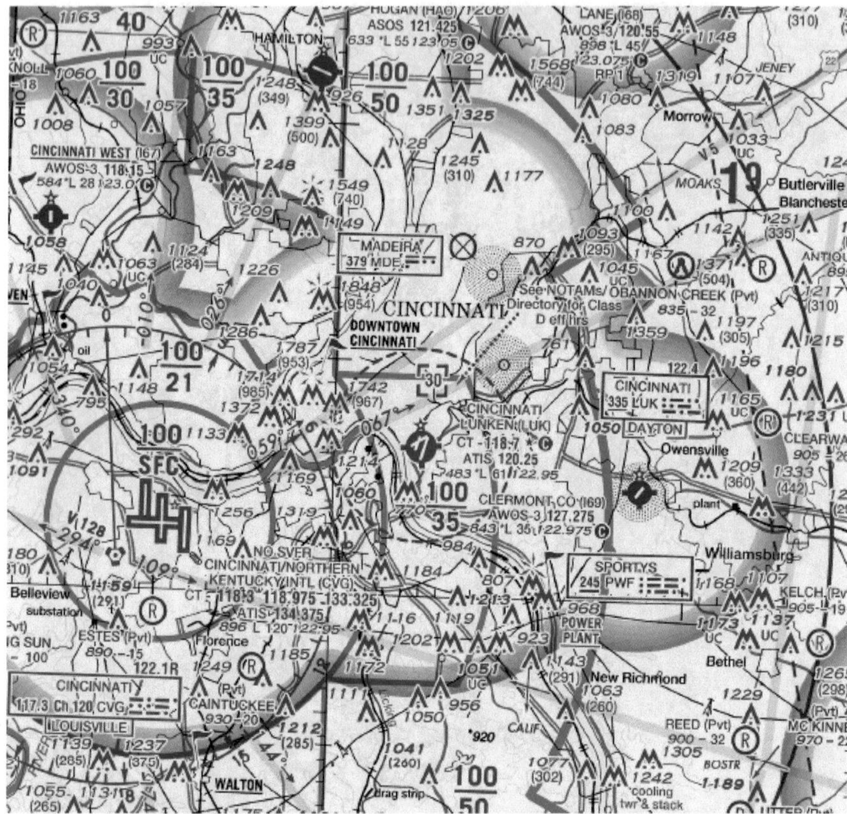

Fig. 2.7 Federal Aviation Administration Chart for Cincinnati, Ohio

determine the relationships (e.g., distance, proximity, and so forth) between objects or spaces within the grid. As will be detailed in the next chapter, this type of grid helps facilitate formalized mechanisms of georeferencing.

Space becomes *place* when "it acquires definition and meaning" (Tuan, 1977, p. 136). Thus, a place is an area that is distinguished or conspicuously separated from other spaces or areas (Stewart, 1975). For example, a *place-name* is a word or series of words that are used to identify physical or administrative features on the earth, its sea floors, or structures located on other planets (Randall, 2001). Assigning place-names is a simple form of georeferencing, and as a result, the vast majority of the earth is ascribed with contextual information pertaining to place. Although these place-names can change over time, often the result of cultural shifts and the political, economic, and sociological processes related to the re-naming process, the embedded georeferences pertaining to place are relatively persistent.

The importance of space and place are not to be underestimated. It is not unreasonable, for example, to suggest that about 50% of all information contains some type of evidence that allows for the process of georeferencing to occur. For instance,

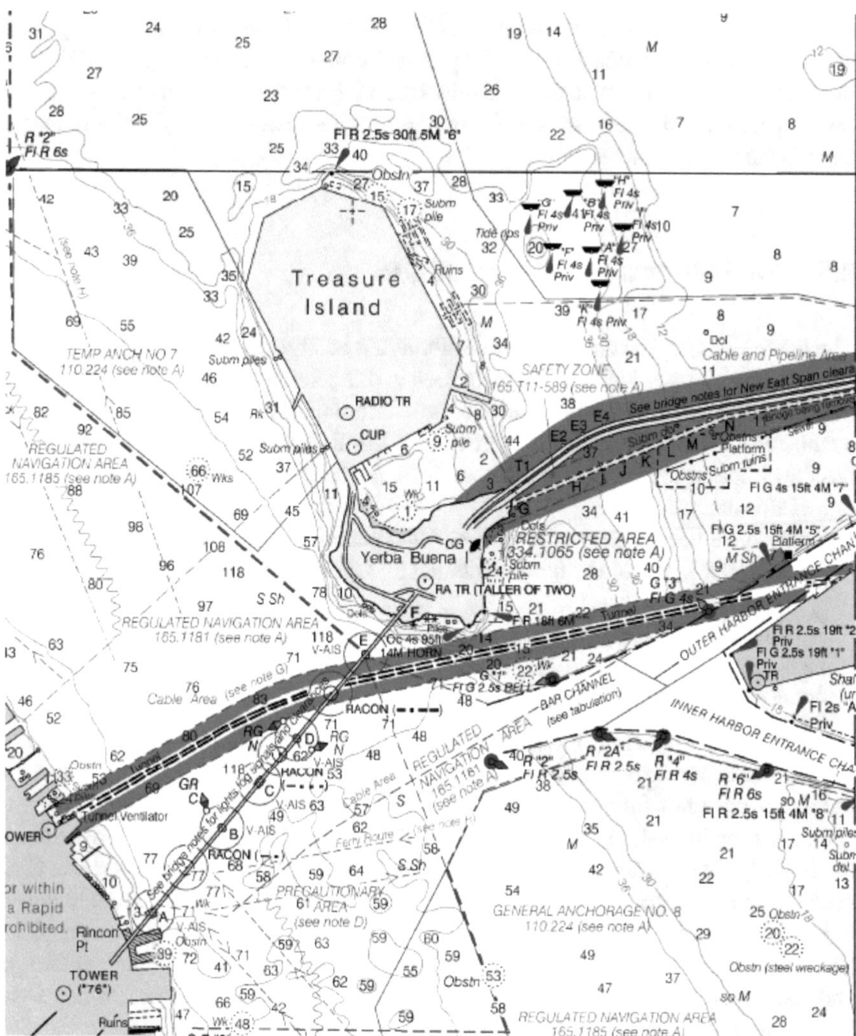

Fig. 2.8 Nautical Chart for the San Francisco Bay Area

a recent study explored five million library catalog records (1968–2000) from the University of California and determined that about 50% of them had one or more place-related subject headings or codes (Petras, 2004). In fact, recent research suggests that nearly 48% of all queries to a library reference service were location-based questions (Bishop, 2011). Thus, it is clear that both space and place are extremely important to humans and absolutely critical in the development of the analytical strategies humans use to organize, process, and reason about their environs. Although the prevalence of place-related information is not surprising, the sheer volume of place-based references is notable. All governments base governance on political boundaries; therefore, all government data retain geospatial locations (Agnew, 2005). Given the enormity of both formal and informal GI available for consumption, the need for a tool to capture, store, manipulate, analyze, manage, and visualize these data emerged.

2.5 An Abbreviated History of GIS

The history of GIS is a complex one. Although it is largely dominated by a few key events and players, which will be recounted in this section, the history of GIS is also littered with many minor innovations, failed companies, and ideas, serendipitous events, and a cast of characters worthy of a Lifetime movie. Although we cannot possibly hope to cover every detail here, readers are referred to Foresman (1998) for more information.

There have been many definitions of GIS floated in the literature over the years (Maguire, 1991), one of the best is detailed by Dueker (1979, p. 106):

> *a special case of information systems where the database consists of observations on spatially distributed features, activities, or events, which are definable in space as points, lines, or areas. A GIS manipulates data about these points, lines, and areas to retrieve data for ad hoc queries and analyses.*

As noted by Maguire (1991); Dueker's (1979) definition tends to emphasize GIS as a database-centric technology. However, it is also important to note that Dueker's definition does not fully capture the frequency with which GIS can be used as a decision support system or as a tool for visualization. It also completely overlooks the importance of raster-based data for many GIS operations and analyses. As GIS packages continue to "do more" with each software release, one simple analogy for defining GIS that students often find helpful is as follows—if one considers productivity software (e.g., Microsoft Word) as central to processing, manipulating, and visualizing textual information, then a GIS (e.g., ArcGIS) is central to processing, manipulating, and visualizing GI.

The first GIS to be developed was the Canada Geographic Information System (CGIS), which was spearheaded by Roger Tomlinson, but involved many others, including Lee Pratt, Head of the Canada Land Inventory, a large team of programmers and computer scientists from IBM, and many other contributors that helped convert remotely scanned data to topologically coded map formats, developed algorithms for error correction and updating, line smoothing and generalization, and many other key components to the CGIS. The reason that CGIS is so important, aside from being the first viable enterprise GIS, is that it spawned an amazing array of cartographic technologies (including the first high precision 48 × 48 in. free-cursor digitizing table), a large-format digital scanner, and the many algorithms and command language components for getting the CGIS to function (Tomlinson,

2012). All of these technologies served as a foundation for subsequent developments in GIS and mapping.

It is also important to acknowledge the conceptual contributions that fueled the GIS revolution, many of which did not directly benefit from advances in computing technology and database development. Although the idea of map overlays had existed for some time (Gardner, Griffith, Harness, & Larcom, 1838), when Ian McHarg, a landscape architect, published *Design with Nature* (McHarg, 1951), he detailed an overlay method that helped display a large range of GI for a given location. In this case, McHarg demonstrated his approach using a road construction project in Staten Island, New York. The engineers charged with building the Richmond Parkway had determined the most cost efficient route would require construction along the Staten Island Greenbelt, a beloved, contiguous system of public parkland, greenspace, and natural areas running through the middle of the island. McHarg provided an alternative path for the road by creating a series of map transparencies that displayed the relative social value of the surrounding landscape. This included information on forest systems, wildlife, residential developments, water, and so forth. On the map transparencies, darker shades of gray suggested more social value, while lighter shades suggested less. Once this information was compared with local geologic information and known hazards, a less intrusive path for the road was determined. Today, the route in question is known as the Korean Veterans Memorial Highway, but the proposed extension analyzed by McHarg, which would have extended the parkway from Richmond Avenue to the Staten Island Expressway, was never built. More importantly, the overlay concept developed by McHarg to evaluate this planning challenge became absolutely fundamental to GIS and remains a core data manipulation, visualization, and processing technique for most (if not all) commercial GIS packages.

In 1965, shortly after the CGIS commenced development, Howard Fisher founded the Laboratory for Computer Graphics and Spatial Analysis (LCGSA) at the Harvard University Graduate School of Design. Aided by considerable funding from the Ford Foundation (Chrisman, 1998), Fisher's goal was to build upon the prototype of the Synagraphic Mapping System (SYMAP) that he developed at Northwestern University. In time, and with help from an all-star roster of collaborators (e.g., Brian Berry, Waldo Tobler, Nick Chrisman) the Harvard Lab helped build a range of foundational tools such as CALFORM, SYMVU (3D views), and GRID (raster processing) and ODYSSEY—a modular system to process and visualize geographic information in vector format. One can also draw a direct line from the LCGSA to Redlands, CA, home to the Environmental Systems Research Institute (Esri) and its founder Jack Dangermond. Dangermond was a graduate student in landscape architecture at Harvard in the late 1960s, working in the LCGSA. Today, Esri is the largest GIS company in the world and Dangermond's net worth is estimated to be $3 billion (Forbes, 2016).

Also in the late 1960s, the U.S. Census Bureau was working on approaches for converting analog maps into numerical renderings from the 1967 pretest of mailout/mailback forms conducted in New Haven, Connecticut. Unfortunately, they found the process to be laden with redundant operations (U. S. Census Bureau, 2015). James Corbett, a Census Bureau mathematician, helped resolve these redundancies by introducing the Dual Independent Map Encoding (DIME) system. DIME was structured using the basic principles of map topology, where intersections, streets, and blocks were encoded as points, lines, and polygons. The DIME system became a key ingredient in the development of the Topologically Integrated Geographic Encoding and Referencing (TIGER) system for the 1990 Census (U. S. Census Bureau, 2015).

As detailed by Longley, Goodchild, Maguire, and Rhind (2015), there were many additional developments in remote sensing technology, satellite data collection such as Landsat, global positioning systems (GPS), and other military technologies that provided key innovations to an ever broadening GIS industry. However, once costs computing hardware began its precipitous decline in the early 1980s and the quality of that hardware began to improve and become more powerful, GIS technology became more widely available and saw increasing use. For example, consider ESRI's ArcInfo 1.0, which debuted in 1982 as the first commercially available GIS software package for mainframe computers. ArcInfo's features were accessed through command lines and included two basic modules: (1) Arc, which included tools for GI input, processing, and visualization, and, (2) Info, a relational database system. The system was not particularly user-friendly, but its full suite of GIS tools were certainly welcomed by users. By the mid-1980s, a version of ArcInfo that ran on IBM compatible personal computers was released, but it was not until the early 1990s that GIS truly moved away from the mainframe and to the desktop. In 1991, ESRI released ArcView 1.0, which for all intents and purposes was a "viewer" for output generated in ArcInfo. A more competitive landscape also emerged in the GIS industry, including MapInfo and their MapInfo Professional software, Strategic Mapping's Atlas GIS package, and Caliper Corporation's Maptitude.

With methods, software, and data all widely available for expanding the GIS user base and broader community, it was becoming more important to standardize data and enable sharing across platforms. For example, at one point in the early 1990s, the vast majority of GIS packages, including Atlas GIS, MapInfo Professional, used proprietary data formats. This was true in the raster-GIS world too, with companies such as LizardTech GeoExpress package promoting their image compression format known as MrSID. Not surprisingly, this stunted industry growth and had a relatively negative impact on GIS usability. In 1994, the Open GIS Consortium was founded to encourage the development and implementation of open standards in geoprocessing. In the same year, the U.S. government weighed in with GI policy of the National Spatial Data Infrastructure (NSDI). From the 2000s onward, with the rise of the Internet and Web 2.0, GIS is more regularly folded into related domains, such as location-based services, the Geoweb, and satellite navigation services. In short, GIS has gone through many changes over the years and will continue to evolve as the demand and consumption of GI increases.

References

Agnew, J. (2005). Sovereignty regimes: Territoriality and state authority in contemporary world politics. *Annals of the Association of American Geographers, 95*(1), 437–461.

Bishop, B. W. (2011). Location-based questions and local knowledge. *Journal of the American Society for Information Science and Technology, 62*(8), 1594–1603. doi:10.1002/asi.21561.

Chrisman, N. R. (1998). Academic origins of GIS. In T. W. Foresman (Ed.), *The history of geographic information systems: Perspectives from the pioneers* (pp. 33–43). Upper Saddle River, NJ: Prentice Hall PTR.

Dueker, K. J. (1979). Land resource information systems: A review of fifteen years experience. *GeoProcessing, 1*, 105–128.

Dutka, J. (1993). Eratosthenes' measurement of the earth reconsidered. *Archive for History of Exact Sciences, 46*(1), 55–66.

Eratosthenes. (2010). *Eratosthenes' geography*. (D. W. Roller, Trans.). Princeton, NJ: Princeton University Press.

Fischer, I. (1975). Another look at Eratosthenes' and Posidonius' determinations of the Earth's circumference. *Quarterly Journal of the Royal Astronomical Society, 16*, 152–167.

Forbes. (2016). Profiles of the World's Billionaires. *Jack Dangermond*. Retrieved from http://www.forbes.com/billionaires/list/#version:static_search:dangermond.

Foresman, T. W. (1998). *The history of geographic information systems: Perspectives from the pioneers*. Upper Saddle River, NJ: Prentice Hall.

Gardner, J., Griffith, R. J., Harness, H. D., & Larcom, T. A. (1838). *Irish Railway Commission*. (Atlas to accompany 2nd. Report of the Railway Commissioners). London.

Kimmerling, A. J., Buckley, A. R., Muehrcke, P. C., & Muehrcke, J. O. (2011). *Map use: Reading, analysis, interpretation* (7th ed.). Redlands, CA: Esri Press Academic.

Longley, P. A., Goodchild, M. F., Maguire, D. J., & Rhind, D. W. (2015). *Geographic information science and systems*. Chichester, England: Wiley.

Maguire, D. J. (1991). An overview and definition of GIS. *Geographical Information Systems: Principles and Applications, 1*, 9–20.

Martin, D. (2005). Socioeconomic geocomputation and e-social science. *Transactions in GIS, 9*(1), 1–3.

McHarg, I. L. (1951). *Design with nature*. New York: Doubleday/Natural History Press.

Oxford Dictionaries. (2016). *American English Definition*. Retrieved from http://www.oxforddictionaries.com/definition/english/space3.

Petras, V. (2004). Statistical analysis of geographic and language clues in the MARC Record (Technical report for the "Going Places in the Catalog: Improved Geographical Access" project, supported by the IMLS National Leadership Grant for Libraries, Award LG-02-02-0035-02. ed.): University of California at Berkeley.

Randall, R. R. (2001). *Place names: How they define the world—And more*. Lanham, MD: Scarecrow Press.

Robinson, A. H., Morrison, J. L., Muehrcke, P. C., Kimmerling, A. J., & Guptill, S. C. (1995). *Elements of cartography* (6th ed.). New York: Wiley.

Stewart, G. R. (1975). *Names on the globe*. New York: Oxford University Press.

Tomlinson, R. (2012). Origins of the Canada geographic information system. *ArcNews*. Retrieved from http://tinyurl.com/nevtnyg.

Tuan, Y. (1977). *Space and place*. Minneapolis, MN: University of Minnesota Press.

U. S. Census Bureau. (2015). Dual independent map encoding. *History*. Retrieved from https://www.census.gov/history/www/innovations/technology/dual_independent_map_encoding.html.

Utrilla, P., Mazo, C., Sopena, M. C., Martínez-Bea, M., & Domingo, R. (2009). A palaeolithic map from 13,660 calBP: Engraved stone blocks from the Late Magdalenian in Abauntz Cave (Navarra, Spain). *Journal of Human Evolution, 57*(2), 99–111.

Chapter 3
0°: A Primer on Geographic Representation

Abstract The purpose of this chapter is to provide a technically accessible overview of geographic representation, geodesy, projections, and georeferencing. All four of these domains function as key building blocks for better understanding the important details associated with calculating distance, direction, and elevation, as well as locating features on the earth's surface and transforming these features into positions on a flat map (or screen) through the use of projections. More importantly, as digital representations of the earth's surface continue to grow in popularity and use, it is also important to provide a primer on how these data are conceptualized and handled in an electronic environment.

3.1 Digital Representation

There are many ways in which the earth, its objects, and/or features can be represented. As detailed in Chap. 2, the Paleolithic map discovered in Spain depicted the entrance to the cave in which it was found and included a series of wavy lines to represent a river, and other animals and features are drawn in relation to the cave (Utrilla, Mazo, Sopena, Martínez-Bea, & Domingo, 2009). Although this representation is more akin to a map, a photograph of the area, which is a two-dimensional representation of light (emitted or reflected) by the earth and its associated objects, would also qualify as a valid representation. The use of text, spoken words, numbers for representing distance, elevation, or temperature also qualify as representations. As detailed by Longley, Goodchild, Maguire, and Rhind (2005), the ability to both conceptualize and formalize such representations allows humans to assemble significantly more information and knowledge about the earth than would be possible by a single individual.

Digital representation consists of complex, coded patterns of zeros or ones (0,1) for representing values or attributes. For example, consider the American Standard Code for Information Interchange (ASCII), which is the most common format for digital text files in computers and on the Internet. In an extended ASCII file, each character is represented by an 8-bit binary number that consists of zeros or ones. For instance, although the text you are reading in this sentence was written in a word processing program, the underlying digital representation of each character corresponds to a series of zeros and ones. For example, the phrase "hello world", without the quotes, translates to "01101000 01100101 01101100 01101100 01101111 00100000 01110111 01101111 01110010

W. Bishop, T.H. Grubesic, *Geographic Information*, Springer Geography,
DOI 10.1007/978-3-319-22789-4_3

01101100 01100100". Although this may seem unwieldy and inefficient, where ten characters are represented by 88 zeros and ones, computers are specifically designed to interpret this information quickly and represent it accordingly.

Thus, the power of this digital representation scheme is rooted in the ability to take almost any information, standardize it through binary formatting and then process and represent it in a myriad of ways. For instance, these binary data can be used to compose digital photographs, generate digital music, send an email or video over the Internet, or digitally represent geographic features in a geographic information system (GIS). In short, the ability of computers and computer networks to handle this information, independent of context or underlying meaning, enables digital data to be copied exactly, stored *en masse*, transferred in bulk at the speed of light, processed efficiently, and analyzed liberally. More importantly, digital data are relatively durable. As will be discussed in Chap. 9 of this book, although the physical deterioration of digital data can occur, it is generally related to the deterioration of the physical media on which it is stored.

3.2 Geographic Representation

Geographic representation is concerned with objects, features, and interactions that occur at the earth's surface or near the surface. As mentioned in Chap. 1, because these features and objects are inherently complex, we are forced to create abstractions, which help simplify complex information for representation, transformation, analysis, and visualization. Longley et al. (2005) note that geographic data are built from "atomic" elements, which are basic facts about the geographic world. The most primitive form of this information would link place and time with some descriptive property or attribute. For example, consider the statement "a house located at latitude 42.322 and longitude −122.825 was burgled on December 23, 1997." This statement ties location, time, and an attribute together. In this instance, the attribute could vary. For example, one could have measured the temperature at this location and time, rather than documenting an event that occurred. One could also vary the location and time, capturing events from a different place, on a different day, or perhaps a different year. Regardless of the underlying details, the ability to produce and abstract facts into their primitive atoms is an important process for geographic representation.

One of the more challenging aspects of dealing with these atomistic representations of geographic data is understanding the differences between attribute types. For readers that have had an introductory course in statistics, many of these terms will be familiar. *Nominal* data correspond to names or identification numbers. These data are explicitly structured to differentiate a particular instance or event from other instances belonging to the same class. For example, the name Comice Drive corresponds to a street in Medford, Oregon, but it is only one of several thousand streets within that city. Nevertheless, we differentiate these streets using these nominal labels. Land use codes and/or types (e.g., residential, commercial, industrial) would be another example of nominal data. *Ordinal* data correspond to attributes that have some type of natural order. For example, if ten neighborhoods are ranked by their crime rates, highest (1) to lowest (10), these rankings would be considered ordinal. *Interval* data conform measurements where the difference between two values is meaningful, but there is no clear definition associated with 0. For example, the difference between 110° and 115° Fahrenheit, is the same as the difference between 0° and 5° Fahrenheit. *Ratio* data have a clearly defined zero value. For example, attributes such as elevation and weight are ratio data. Where elevation is concerned, we know that a value of zero corresponds to sea level. We also know that an entity with zero weight is, in fact, weightless. Lastly, there are *cyclic* geographic data (Chrisman, 1997). Cyclic data include information on flows and their compass direction. As detailed by Chrisman (1997), dealing with these data are difficult because a compass consists of 360°, but the number that follows 359 is 0. As a result, attempting to perform basic mathematical operations on these data, such as averages, is difficult. For more details on the analysis of circular and/or directional data, see Fisher (1995) and Brunsdon and Corcoran (2006).

Considering the sheer number of features on the earth's surface and their associated attributes, atomistic representations require basic organizing principles to be effective. As detailed in Chap. 1, there are two basic data models for handling the abstraction of geographic information (GI). First, the object view considers GI as a series of discrete entities located in geographic space. Compositionally, this includes information pertaining to the locations and shapes of geographic objects (e.g., points, lines, and areas) and entities (e.g., combinations of points, lines, and areas) and the relationship(s) between them (Longley et al., 2005; O'Sullivan & Unwin, 2010). The compositional facets are usually stored as (1) coordinates (e.g., latitude and longitude), that function as absolute measures of geographic location, and (2) topology, which is the way in which geographic features are interrelated or arranged. Figure 3.1 is fairly self-explanatory, illustrating an example set of entities for an object view.

The field view considers GI to consist of properties that continuously vary across space. For example, elevation above sea level is a continuous geographic variable within the field view. Binary classifications (e.g., 0,1) for raster-based field representations are also possible, where cells have a value of 1 when certain conditions are present (e.g., forest) and 0 otherwise (O'Sullivan & Unwin, 2010). Figure 3.2 illustrates several examples of a region conceptualized using the field view. Specifically, Fig. 3.2a is a digital elevation model for the area in/around Cave Creek, Arizona, with Black Mountain serving as the focal point. In this case, elevation is continuous, represented by series of raster cells. Figure 3.2b c retain the same underlying raster structure, but illustrate a shaded relief map (i.e., hillshade) and a solar insolation map, respectively, for the Cave Creek area. Lastly, Fig. 3.2d departs from the use of raster data and instead displays a contour representation of the surface illustrated in Fig. 3.2a (elevation). Here, polylines are used to represent (and connect) interpolated positions between grid cell centers to identify lines of constant elevation. In this instance, the contour intervals are 10 ft.

The point to be made here, no pun intended, is that both the object and field view have been designed to handle, as best they can, vast quantities of information regarding the earth's surface. We know that information is lost in the process, but we hope that

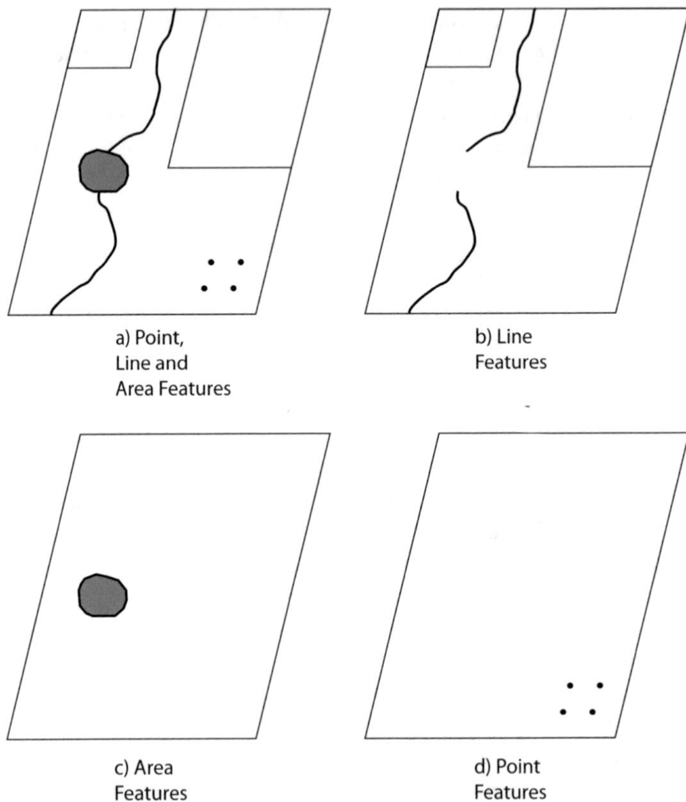

Fig. 3.1 Point, line and area features on a map

enough information is preserved, with sufficient accuracy and precision, to provide a good abstraction of real-world phenomena. To accomplish this task, the object view relies upon vector data structures, and the field view relies upon raster data structures. Both types of representations have their advantages and disadvantages. As illustrated in Fig. 3.2d, fields can be represented using vector structures, and objects can be represented using rasters (not illustrated). Although *interchangeable* is too strong a word here, both raster and vector data can be used to represent geographic variation quite efficiently, but their relative strengths and weaknesses must be considered.

3.2.1 Vector

At the heart of vector representation, all entities and objects are represented as points, lines, and polygons, but the underlayment for all lines and polygons are *points*. Specifically, the vector data model uses points, stored as real-world geographic coordinates, to build more complex objects, such as lines and polygons. As a result,

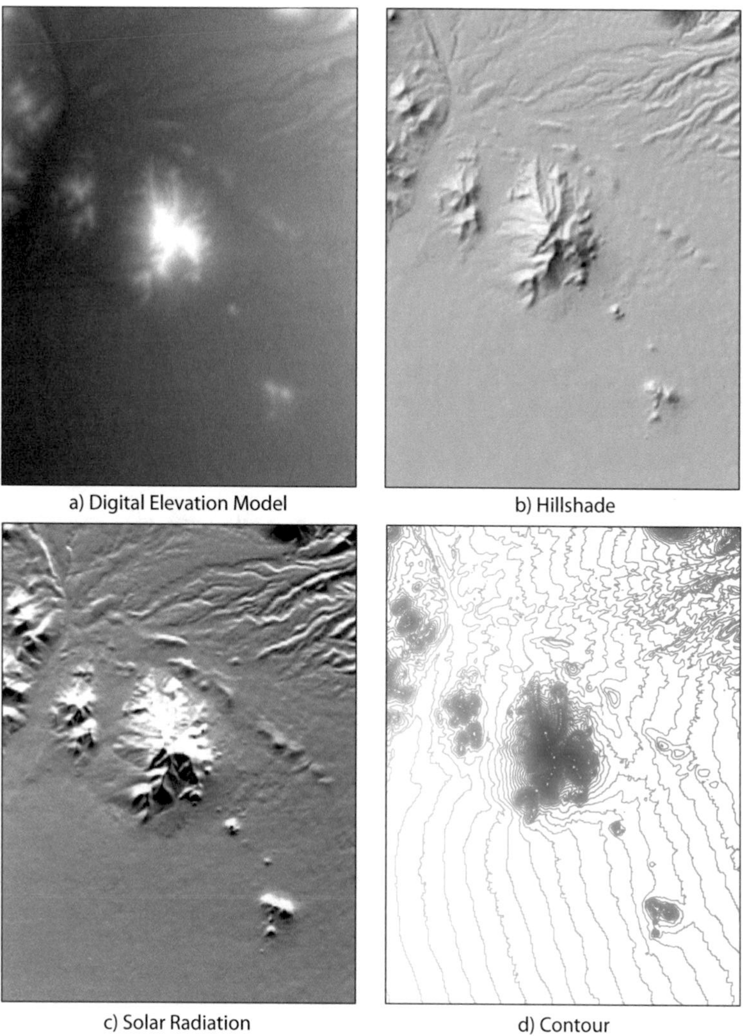

Fig. 3.2 Raster Representations and an alternative vector visualization

all of the objects in a map that uses vector representation can be treated as a graph. Without delving too deeply into the mathematics (see (Harary, 1969) for more details), a graph (*G*) is defined by two sets, *V* and *E*. *V* is a finite set of points (i.e., vertices) and *E* is a set of edges (i.e., arcs or lines). *G* also consists of incidence relations, $V \times V$; that describe which vertices are connected by edges. When the arcs are directed, *G* is a directed graph. When the arcs intersect only at the vertices, *G* is a planar graph. The regions defined by a planar graph are called *faces,* and the remaining unbounded region is referred to as the exterior face.

Simply put, nodes are used to represent point features, such as an antenna tower. Nodes are also used to represent the beginning and ending of all linear features in

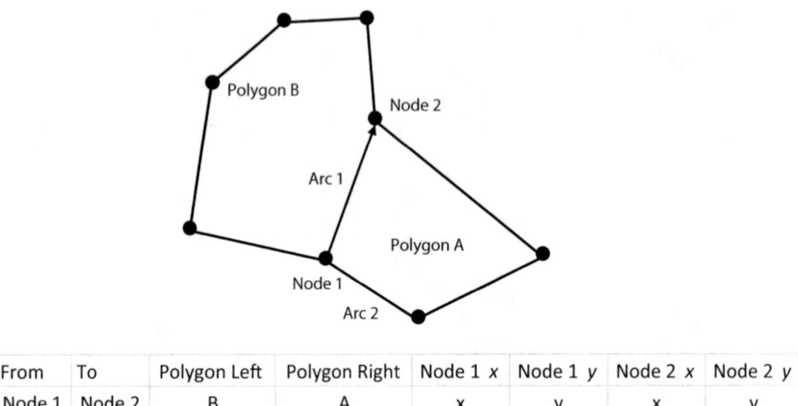

Arc	From	To	Polygon Left	Polygon Right	Node 1 *x*	Node 1 *y*	Node 2 *x*	Node 2 *y*
1	Node 1	Node 2	B	A	x	y	x	y

Fig. 3.3 Vector topology

the graph, as well as appearing at the intersection of all linear features such as roads and pipelines. Nodes and arcs are also the building blocks for all polygons, and these bounded regions can correspond to geophysical features such as lakes, or administrative features such as Census tracts.

There are several advantages in using vector representations. First, data can be represented in their original resolution and their original form, with very little need for generalization. By reducing the need for severe abstraction, these data are more intuitive and aesthetically pleasing. Second, when using vector representations, one only needs to specify the locations of the nodes that form the vertices of a polygon to define an object. In fact, when this is done *en masse*, and the relationships between nodes, lines, and polygons are specified, topological relationships can be defined, built and stored. These topological relationships are important because they provide us with the "…properties which define relative relationships between spatial elements… including adjacency, connectivity, and containment" (McDonnell & Kemp, 1995, p. 88). Figure 3.3 illustrates a basic suite of topological relationships that can be found in vector representations. In this particular case, we have eight nodes (only Nodes 1 and 2 are labeled), two polygons (A and B), and nine arcs (only Arc 1 is labeled). Each node is defined by a coordinate pair (*x*,*y*), which would consist of a latitude and longitude in the real-world. Arc 1 is directional, starting from Node 1 and ending at Node 2. Polygon B is located to the left of Arc 1 and Polygon A is located to the right of Arc 1. In a nutshell, this simple set of topological relationships allows us to say many things about the represented objects. For example, we know that Polygon A is adjacent to Polygon B. We know that Arc 1 forms a border for both Polygons and that it is connected to Arc 2. If we were to add a point in the middle of Polygon A, we could make a definitive statement regarding containment of that new point. All GIS packages that can read, interpret, and display vector data make use of these relationships.

One major disadvantage of vector representations is that they are not particularly good at handling continuous coverages (e.g., elevation). As illustrated in Fig. 3.2d, substantial data generalization (via interpolation, in this case) is required. This adds uncertainty to the data and the probability for misrepresentation increases substan-

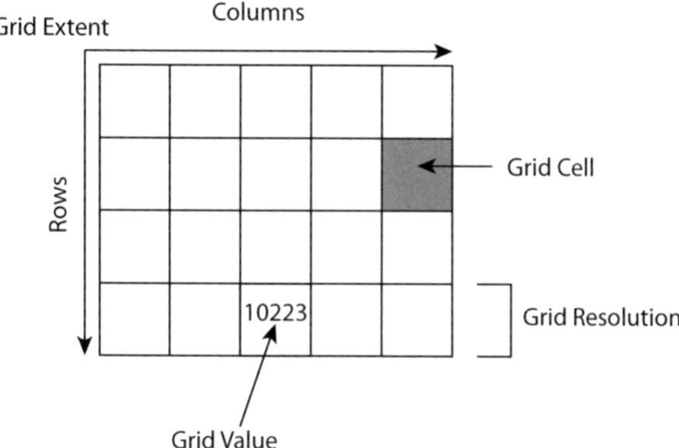

Fig. 3.4 Raster basics

tially. A second disadvantage is that the drawing, manipulation, and analysis of dense vector layers can be computationally intensive, effectively limiting their functionality when a massive number of objects exist in a geographic base file.

3.2.2 *Raster*

Unlike vector, raster representations do not describe discrete geographic entities or objects. Instead, raster data are used to record and represent spatial variation by capturing the locational pattern of an attribute for a region. The spatial interval for capturing variation is typically regular, consisting of an array of cells that are square or rectangular, but there are other geometric tessellations (e.g., triangular or hexagonal) than can also be implemented. Once the study region is tessellated, all of the spatial variation associated with an attribute is then expressed by assigning the appropriate properties to the regularized cells. Raster data include aerial photographs, satellite imagery of all types, and well as images captured by drones.

Figure 3.4 illustrates a generic raster data structure. One major advantage of using raster representations is that very little spatial referencing information is required for a grid. Unlike vector systems, where all points and polygons require explicit geographic coordinates for representation, raster grids only require a single origin. Based on the explicit GI associated with the origin and the uniform offsets associated with each grid cell, the specific location of any given cell can be determined easily. Raster grids have a number of additional, basic properties associated worth detailing. First, the *grid extent* is simply a measure of the overall size of the grid. For Fig. 3.4, the grid extent is 4 × 5, but we could easily relabel this with specific distances (e.g., 40 m × 50 m). Grids have a resolution. In this case, each cell in Fig. 3.4 corresponds to 10 m^2. All grid cells have a value, which corresponds to the attribute being mapped. The value in Fig. 3.4 corresponds to elevation. Finally, grids are generally organized via rows and columns, but there are many alternative ways to organize and store raster data. For more details on topics such as run length encoding, quadtrees, or range trees, see Chang (2006).

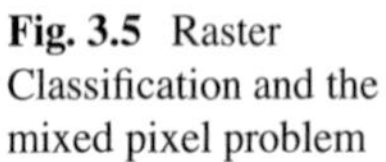

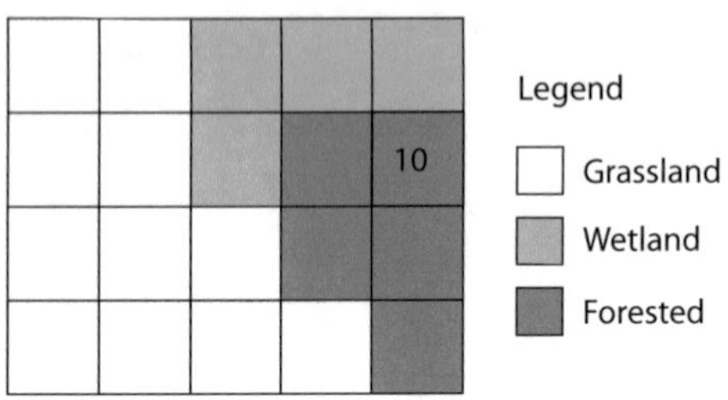

Fig. 3.5 Raster Classification and the mixed pixel problem

Another advantage of raster representations is the relative simplicity in which it stores attribute information. For example, Fig. 3.5 illustrates a raster representation of land cover with each shade of gray representing a nominal classification (e.g., grassland, wetland, and so forth). Because each cell accommodates only a single attribute, data analysis is both easy and fast to perform. It is also relatively simple to integrate nominal, classified data with GI on continuous coverages, such as elevation or temperature when using raster data.

One of the major limitations associated with using raster representations is that intra-cell variation is lost when measuring attributes. For example, Cell 10 in Fig. 3.5 might consist of both forested and wetland areas, especially as one nears the upper edge of the pixel. This is known as the "mixed pixel problem" (Fisher, 1997, p. 680). Under such conditions, one must select a rule to apply for consistently labeling pixels that might fit more than one classification. For instance, one could label each pixel based on the largest share rule, where the dominant characteristic (e.g., forested) wins the label. Other rules include labels based on the central point of the cell or some type of weighted average algorithm. Regardless of the method chosen, it must be applied consistently. A second disadvantage is that pixel size determines the resolution at which the data are represented. If an analyst needs something more resolute, it may not always be available. Third, it is extremely difficult to use raster systems for capturing, representing, and analyzing network data, especially if pixel sizes are fairly large. Figure 3.6 illustrates the relative frailty of raster representations for linear and/or network data. In this instance, the pixel size is far too big for accurately capturing the path of the river. In addition, Fig. 3.6 does a good job in illustrating one last problem for both raster and vector data. Conversion between the two representation systems can be clunky and inaccurate, especially with network data.

3.3 Geodesy

Geodesy is the science of measuring the size and shape of the Earth. As detailed in Chap. 2, Eratosthenes was able to determine that the earth was spherical by comparing the geometrical relationships between Alexandria and a well located on the Tropic of Cancer in Syene. Today, we know that the earth is not a perfect sphere. It is more ellipsoidal in shape. More specifically, the earth is an oblate ellipsoid, which means it is slightly flatter (i.e., oblate) at the top and bottom, and rounder on the

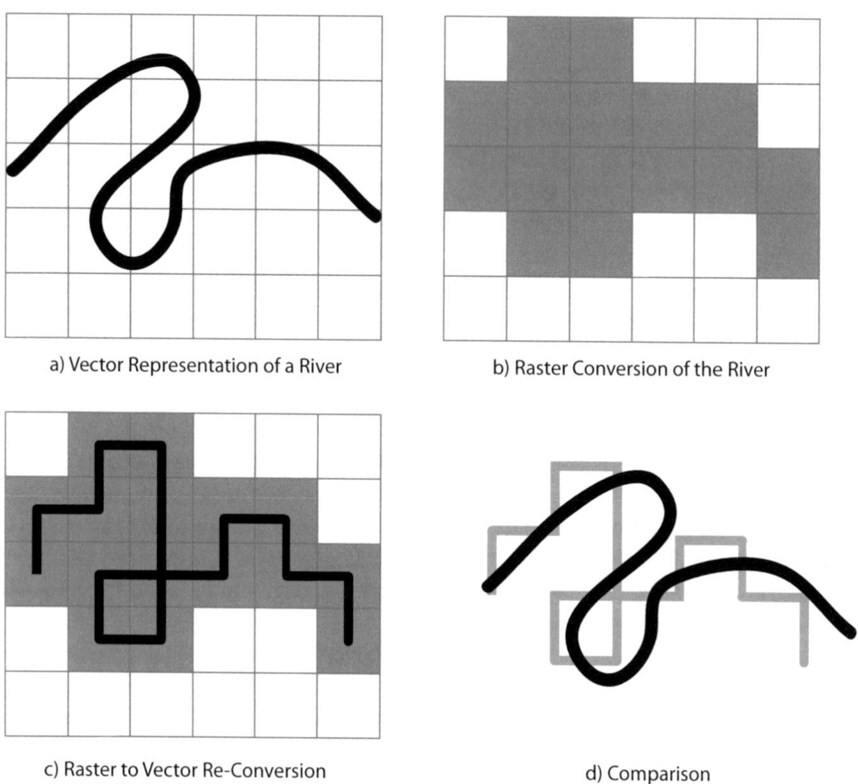

Fig. 3.6 Vector to Raster to vector conversion problems

sides. Phrased somewhat differently, the circumference of Earth is larger at the equator than it is from pole to pole. This general shape can be mimicked if one takes a tennis ball and flattens it down slightly at the top and bottom.

Not surprisingly, there are variations in the mathematical measurements and associated ratios used to capture the degree to which the Earth is an ellipsoid. For example, the World Geodetic System (WGS) measurements from 1972, which were determined from satellite data, suggest that the equatorial radius was 6,378,135 m, while the polar radius was 6,356,750.5 m. WGS 1984 suggests that the equatorial radius is 6,378,137 while the polar radius is 6,356,752.3 m. Today, the WGS 84 ellipsoid is used as the global standard for mapping.

It is also important to remember that the earth is a collection of materials in constant flux. In part, this can be attributed to irregularities in its magnetic field, but also to an active, liquid core that is constantly moving, shaking things up, and spilling things out. To better account for these subtleties, the earth can be represented as a geoid. This is a three-dimensional shape that is approximated by a sea-level, equipotential surface (Robinson, Morrison, Muehrcke, Kimmerling, & Guptill, 1995). In other words, this is a surface where the ocean would be allowed to flow freely across the entire surface of the earth. More importantly, gravity is uniformly equal to its strength at the calculated mean sea level across the entire equipotential surface. Thus, if the earth were devoid of mountain ranges, trenches, and other vertical irregularities, the geoid would be identical to an ellipsoid. However, because the earth is both geologically and topographically complex, the geoid *is* different from the ellipsoid, deviating by as much as 100 m in certain locations.

Pragmatically, these differences do not mean much for casual users of GIS or basic mapping applications. Most commercial systems automatically account for the differences spherical, ellipsoidal, and geoidal approximations of the earth's surface. However, it is important to recognize that these approximations are the building blocks for many of the most important functions in GIS and spatial analysis, including distance, direction, and area measurements. For now, we can parse these differences into a simple, user-friendly framework:

1. The authalic sphere is the best reference surface for small-scale maps that are used to represent hemispheres, continents, or large countries. At this scale, there is no discernable difference between the sphere and ellipsoid. More importantly, as detailed by Robinson et al. (1995), the complexity of map projections increases dramatically when moving from those used for a sphere, to those used for an ellipsoid. This is best avoided, if possible.
2. The ellipsoid is the best reference surface for large-scale maps that are used to represent small areas in topographic maps, metropolitan areas, and the like. For example, the WGS 84 ellipsoid can be used by global positioning systems (GPS) to achieve measurement accuracies to 1 m. Thus, when capturing distance, direction, and area measurements for large-scale regions, one can be quite confident in the results.
3. The geoid is the best reference surface for extremely large-scale maps and land surveying, where each centimeter for horizontal and vertical positions makes a difference. For example, one would use the geoid for creating a cadastral map for a new residential development in a city or to determine the exact footprint of a building in the development.

3.3.1 The Graticule

Although it can seem complicated, the graticule is a clever system for defining locations on the earth's surface. It consists of a grid of horizontal (parallels) and vertical (meridian) lines that cover the earth. Parallels run in an east-west direction and meridians run in a north-south direction. Both sets of lines are equally spaced, with meridians converging at the north and south poles. Hipparchus, the well-known Greek astronomer, geographer, and mathematician developed a numbering system for the graticule called latitude and longitude.

Latitude corresponds to the north-south, angular distance from the equator. The equator is 0° latitude, whereas the north and south poles are 90°. The standard convention for differentiating north-south angular distance is to use the letters N (for north) and S (for south). For example, Phoenix, Arizona is located at 33.45° N. Longitude is a slightly more complex concept because it is associated with an infinite set of meridians that are arranged perpendicularly to parallels. Today, the prime meridian serves as the meridian of reference (0° longitude), but there have been others (Howse, 1980). Longitude corresponds to the angle formed by a line going from the intersection of the prime meridian and equator to the center of the earth. This same line then goes back to the intersection of the equator and the local meridian where the object of interest is situated. Longitude values range from 0° to 180°, and the standard convention for differentiating east-west angular distance is to use the letters E (for east) and W (for west). For example, Phoenix Arizona is located at 112.06° W. When latitude and longitude are combined (33.45° N, 112.06° W), a geographic coordinate is created, and it can be used to pinpoint a location on the earth's surface.

To minimize confusion and in our ongoing effort to make this introductory material as gentle as possible, we have no intention of subjecting readers to a lengthy discussion on the differences between geodetic and/or geocentric latitude/longitudes. In short, geodetic latitude is used for ellipsoidal surfaces, while geocentric latitude is used for spheres. As detailed by Kimmerling, Buckley, Muehrcke, and Muehrcke (2011), although the mathematical formulation for geodetic and geocentric longitudes is different, there is no need to differentiate between the two because the result is nearly identical. For more details and a very technical discussion on these topics, see Bugayevskiy and Snyder (1995).

3.3.2 The Graticule and Distance

There are several additional properties of the graticule worth noting, but all of these details assume the use of an authalic sphere. More specific information for ellipsoidal or geoidal information will be detailed in the next section when the topic of projections is covered.

With this caveat in mind, basic geometry tells us that the shortest distance between two points is a straight line. Figure 3.7 displays the intersection of the sphere with a horizontal plane passing directly through the center. This intersection that divides the earth in half (i.e., hemispheres) is called a great circle. This is the largest possible circle that can be drawn on a sphere. Where the graticule is concerned, the equator is a great circle. When two meridians are combined (east and west), they also form a great circle. All other circles that can be drawn on a sphere are called "small circles".

When considering distance, it is important to remember that the surface of an authalic sphere is curved, so it is extremely difficult to follow a truly straight line because it would require penetrating the earth's surface to move between points, especially if the distance is large. However, it is possible to follow that straight (terranean) line above the earth's surface, exactly, using an arc. Figure 3.7 illustrates the trace of the intersection associated with a plane on the authalic sphere between points A and B. Here, the two points are connected by a great circle (arc), which is the shortest route over the surface of the sphere between any two points. The use of great circles is extremely important when calculating distances, especially if the distances between points are large (e.g., global airline routes).

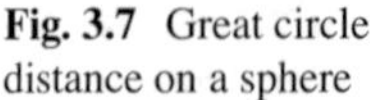

Fig. 3.7 Great circle distance on a sphere

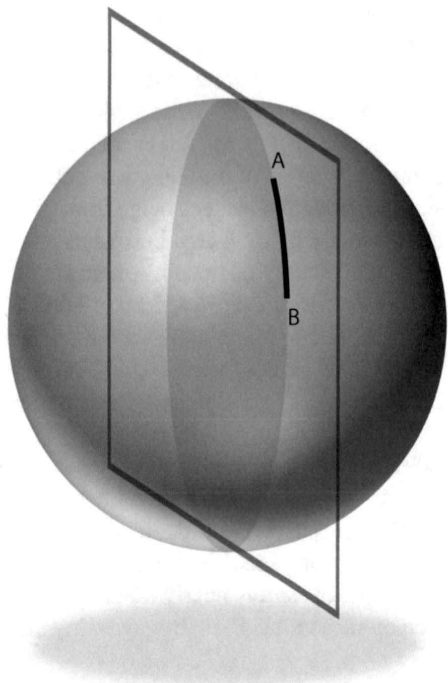

3.3.3 The Graticule and Direction

The mathematical and conceptual foundation of direction is quite complex. By definition, direction refers to a course along which something or someone moves. However, for cartographic purposes, the direction is generally defined as an angular deviation from a baseline (Robinson et al., 1995). As detailed above, since the shape of the earth has been roughly established as a sphere, the north pole was readily adopted as the baseline from which directions could be determined. Interestingly, there are alternatives. For example, magnetic north, as determined by a magnetic compass and its reading of the earth's field of magnetic force was widely adopted for surveying. However, magnetic north is rarely aligned with local meridians (i.e. true north). In fact, magnetic north is relatively dynamic (Olsen & Mandea, 2007), and is currently moving approximately 50 km per year (Olsen et al., 2006). The difference between true north and magnetic north is referred to as magnetic declination and the most current information on the magnetic north pole must be used to determine these differences. Lastly, there is grid north, which refers to the northward direction of grid lines for a given map (typically some type of ordinance/survey map). The general differences for all three types are illustrated in Fig. 3.8 and summarized as follows:

- True north is the direction toward the spherical north pole
- Magnetic north is the direction toward the magnetic dip pole
- Grid north is the direction along vertical grid lines for a map

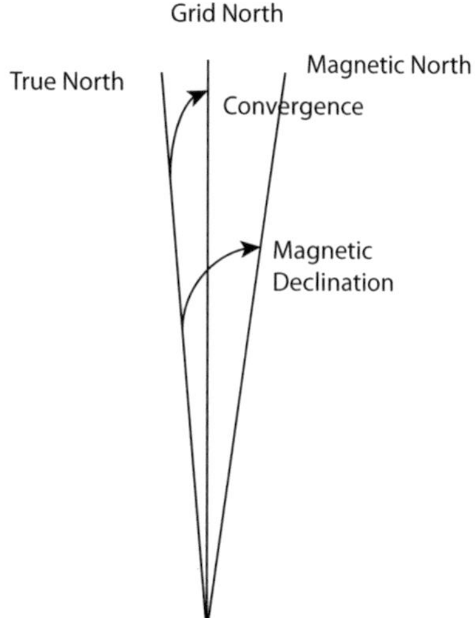

Fig. 3.8 Potential variations in true, grid, and magnetic north

- Magnetic declination is the difference in direction between magnetic north and true north
- Convergence is the difference in direction between grid north and true north.

The reason all of this matters is because when a graticule is in use, the direction of a line is relevant. These lines are generally referred to as a bearing or azimuth. For GIS and mapping more generally, there are two variations of an azimuth that are worth detailing. Azimuth refers to the "horizontal angle measured in degrees clockwise from the north reference line to a direction line" (Kimmerling et al., 2011, p. 256). Values range from 0° to 360°. However, we have already established that the only locations where an azimuth is constant would be along a meridian or an equator. Thus, a true azimuth must make use of the angular calculations with reference to a meridian to obtain an accurate measure of direction. Also relevant is the constant azimuth, or a loxodrome. This is a line that intersects each meridian at the same angle and takes on a spiral shape along a constant path as it approaches true north. When navigating, we already know that the use of a great circle ensures the shortest path between two points on a sphere, but continuous changes in direction to follow these circles is impractical for ships or airplanes on long trips. Instead, navigation charts are made using special map projections and constant azimuths. As a result, the great circles appear as straight lines along the graticule, vastly simplifying navigational paths.

3.3.4 *The Graticule and Area*

As will be detailed in the next section, there are many different types of map projections, most of which have a specific purpose or application. Projections are needed because representations of the earth as a sphere (three-dimensional) must be converted into two dimensions to fit onto a screen or analog paper map. The relationship between the graticule and area is relatively easy to visualize, but the underlying mathematics associated with maintaining an accurate representation are somewhat complex. Consider Fig. 3.9, which shows the graticule in 10° increments for north-south and east-west. The quadrilaterals used to represent the graticule,

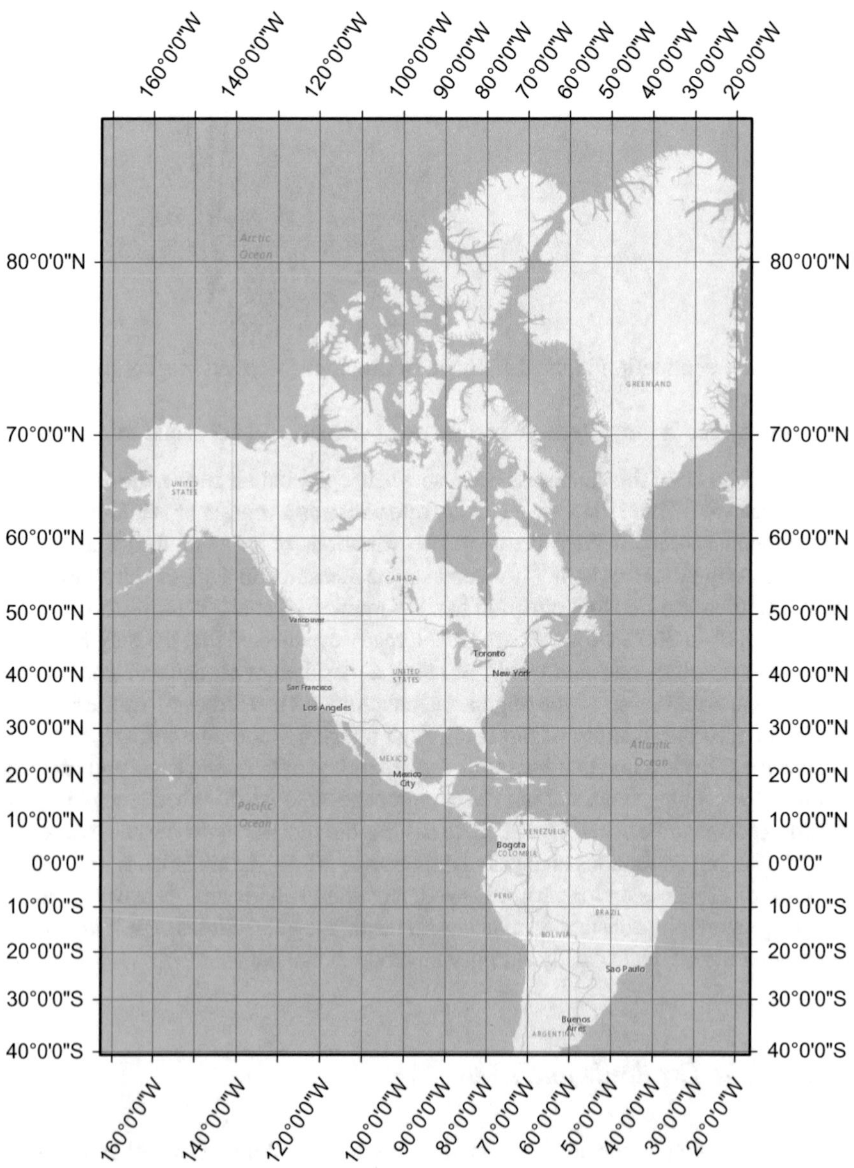

Fig. 3.9 The Graticule

which are bounded by parallels and meridians are regularized, but their relative sizes vary as one moves toward the equator or toward the poles. Specifically, the quadrilaterals decrease in size as one moves toward the poles, but again, this decrease is constant and easily calculated using basic mathematical equations. This matters for cartographic representation because projections can be used to preserve the surface area of features within latitude bands, but without fail, this distorts shapes. Projections can also be formulated for preserving the area for all quadrilaterals on the sphere, but this also distorts shapes. This is why places like Greenland look so large on a map using the Mercator projection.

In the next section, we detail how basic map projections work and why they are important for GI and associated efforts to visualize or analyze them. Further, we provide some basic guidance on which projections should be used for different visualization or spatial analysis tasks.

3.4 Map Projections

Broadly defined, a map projection is the mathematical transformation of the spherical surface of the earth into a planar surface. This process is not without flaws. As detailed by Krygier and Wood, "No map projection preserves the attributes of a globe, which maintains shape, area, distance, and direction. Map projections can preserve one or two of the attributes of the globe, but not all at once" (Krygier & Wood, 2011, p. 84). This quote reiterates that the projection process is imperfect but not so flawed that projected maps lack utility. In fact, quite the opposite is true. With that in mind, it is important to acknowledge the fundamental properties of map projections briefly before key types are detailed.

3.4.1 Projection Properties

Robinson et al. (1995, p. 61) provide an intuitive way for readers to understand the projection process, dividing it into two stages. First, consider a reduced form of the earth—one that has been shrunken, yet retains its spherical shape. This smaller, hypothetical sphere is labeled as a reference globe, and it has the same scale that is desired for the flat map to be created. In turn, this reference globe will be transformed into a flat map, displaying its three-dimensional form on a two-dimensional surface.

Although the basics of map scale were already addressed in Chap. 2, scale assumes a particular significance for map projections. There are two basic types of scale, the *actual scale,* and the *principal scale*. The principle scale is defined by generating a representative fraction (RF) for the reference globe. The RF is easy to calculate. One simply divides the earth's radius by the radius of the reference globe. Thus, the actual scale of the reference globe is identical to the principal scale because the reference globe has not undergone any additional transformations. One basic metric for clarifying the relationship between actual and principal scales is the *scale factor* (SF), which is defined as follows:

$$SF = \frac{Actual\ Scale}{Principal\ Scale} \tag{3.1}$$

By definition, this means that the SF = 1.0 for every location on the reference globe. However, when the globe surface is transformed into a two-dimensional, flat map, the actual scale will vary, as will the SF. Consider for example a reference globe that has a principal scale of 1:100,000,000. When it is transformed into a flat map, one that uses a Mercator projection, the actual scale of the map is going to vary. For instance, at the Equator, SF = 1.0 and the actual scale = 1:100,000,000. This is a line of tangency, where the projected surface "touches" the reference globe. There is no map distortion here. However, as one moves northward, the actual scale begins to vary. For example at 30°N, the SF = 1.15 and the actual scale is 1:86,600,000. This means that the actual scale is larger than the principal scale of the reference globe. This variation grows as one moves northward. The point to be made here is that readers must be cognizant of these differences between principal and actual scales, because they are critically important for understanding the impacts of geometrical transformations between three-dimensional spheres and two-dimensional maps.

A second map projection property worth detailing is *completeness*. Most world maps do not show the entire world. In part, this can be attributed to the inability of transformation equations to be applied to the entire range of latitude and longitude (Kimmerling et al., 2011). For example, on a Mercator projection, the *y*-coordinate for the north pole is equivalent to infinity (*ibid*), which is why most world maps using the Mercator projection stop at 80°. *Continuity* is also an important projection property. Once again, representing an entire sphere on a flat screen or sheet of paper is challenging. As a result, one must decide where to "break" the map for visualization purposes. For most world maps, the break/discontinuous locations appear on the opposite side of the map, even though they are located on the same meridian. As detailed by Kimmerling et al., (2011), this can be a source of confusion for map readers because it strongly violates established proximity relations. For example, it is best to avoid breaking Asia into two, seemingly discontiguous land masses on a map. Instead, best practice dictates that maps of individual continents or large countries should be displayed with no breaks.

3.4.2 Geometric Distortions

Lastly, we are also concerned with the geometric distortions associated with projections. There are four core elements of projections that relate to which properties of the sphere are best preserved. First, the *conformal* property ensures that the shapes of small features on the earth's surface are preserved. Phrased somewhat differently, the angles on the reference globe are preserved, and the scales of the projection for both *x* and *y* coordinates are equal. Conformal maps are used for navigation and as detailed earlier, loxodromes ensure that a straight line drawn on a conformal map will have a constant bearing.

Second, the *equal area* property ensures that all areas drawn on the map will be proportional to those on the reference globe and/or the earth's surface. The mechanics of this are relatively easy to comprehend. As the scale is adjusted along meridians and parallels, when shrinkage occurs in one area, exaggeration in others helps preserve area. As noted by Kimmerling et al. (2011), equal area maps compact, elongate, shear, and skew features on the globe, including the quadrilaterals of a graticule. That said, equal area maps are best used for calculating the size of administrative units (e.g., states, counties, residential lots, etc.) or mapping global geographic phenomena such as population density.

Equidistant maps attempt to preserve spherical great circle distances, but this is impossible to accomplish on a flat map. Consider the requirements. For a map to preserve distance, the scale would need to be equal in all directions and between all locations. We have already established that the actual scale varies on a map, which also means that the great circle distances are distorted. It is possible, however, to generate an equidistant projection that uses a series of lines radiating out from a single point on the map (e.g., north pole). With this type of setup, at the very least, we know the distances are not distorted along those lines because they represent true, great circle distances. Equidistant maps are best used for long distance route planning.

The preservation of direction is accomplished with azimuthal projections. Similar to the limitations of the equidistant maps, there is no single projection that can accurately represent all directions from all locations on the earth's surface. That said, azimuthal projections can be used to create a map where all lines are radiating out of a single point (e.g., north pole) accurately portray direction. Azimuthal maps are also excellent for long distance route planning and navigation.

3.4.3 Projection Surfaces

The three most common projection surfaces (Fig. 3.10) are planar, conic, and cylindrical. Cylindrical projections are analogous to wrapping a piece of cylindrically shaped paper around the earth, projecting the earth's features to it, then unwrapping the paper. Figure 3.10a displays a tangent case for the cylindrical projection, where the scale factor is 1.0 when it touches the globe surface. In this case, the SF is 1.0 at the equator. Planar projections, which are also called azimuthal projections, are analogous to touching the earth with a sheet of flat paper and then projecting the earth's features to it. Figure 3.10b illustrates a secant case of the planar projection, where once again, the scale factor is 1.0 where the plane touches the globe. Finally, conic projections are analogous to wrapping a sheet of paper around the earth that is cone shaped and projecting the earth's features to it. Figure 3.10c illustrates a secant case of the conic projection. These types of maps are best suited for mid-latitude regions where the cone makes contact with the globe and the SF is 1 or close to it. Regardless of the projection chosen, it is important to remember that all three of these projection surfaces can have conformal or equal area properties, but they cannot have both, simultaneously.

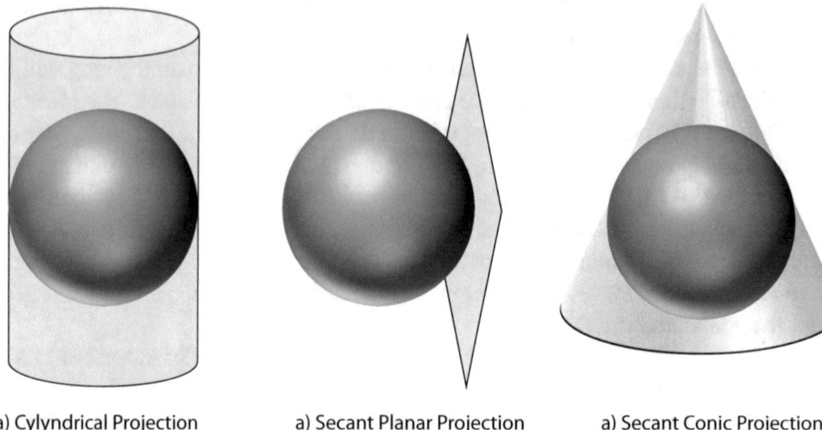

Fig. 3.10 The three major classes of map projections

One last facet on projections worth mentioning for readers (but sparing the details) is the selection of projection aspects. Projection aspect refers to the location of the points of tangency or lines of tangency for the projection surface. Although these can fall anywhere, in theory, projection aspects typically take fairly standard forms. For planar projections, they are labeled equatorial, oblique, and polar (normal). For cylindrical projections, they are labeled equatorial (normal) and transverse. For more details on projection aspect and their impacts on the graticule, see Robinson et al. (1995).

3.4.4 Popular Projections and When to Use Them

There are many different types of map projections. As alluded to earlier, each one has a specific purpose or strength. However, there is also ample opportunity for misapplication and misuse of projections, depending on their preservation properties. We cannot hope to detail all of them in this chapter, but there are four projections that all consumers of GI need to be aware of and use accordingly. Each is detailed below.

3.4.4.1 Cylindrical Equidistant (i.e., Plate Carree)

This is one of the most simple and widely used projections for GI. It is also one of the most dangerous, especially when it comes to spatial analysis. The Plate Carree projection is widely referred to as "unprojected" because it maps longitude as x and latitude as y, resulting in a heavily distorted representation of the earth. It is neither conformal, nor equal area, but the one major advantage of Plate Carree is that it maintains the correct distance between every point on the map and the equator. Many GI are redistributed in this projection, but this has serious implications. Unless users reproject GI into Cartesian coordinates, virtually all calculations pertaining to distance, area, or direction will be inaccurate. For example, consider a basic distance calculation between two, non-equatorial points (*x*,*y*) in a GIS where the layers are represented by a cylindrical equidistant projection. Can this projection, in any way, hope to reflect the required great circle distance for making an accurate assessment of distance? The answer is no. For more details on how this is problematic for spatial analysis, see Murray and Grubesic (2012).

3.4.4.2 Universal Transverse Mercator (UTM)

UTM is a system of projections that are most commonly used with data that has national, global coverage. It is based on the Mercator projection, which is equatorial, but it has been flipped to the transverse (wrapping the cylinder around the poles, rather than the equator). Much like the Mercator projection, UTM is conformal, but the loxodromes are no longer straight lines. In total, there are 60 UTM zones and the major advantage of this organizational structure is that for any given zone, the projected north-south strips (also known as "gores") have almost no shape distortion and very little areal distortion. As a result, spatial analysis within a UTM zone is quite accurate. For example, the distance between points can be calculated with no more than 0.04% error (Longley et al., 2005, p. 121). Many 1:24,000 scale United States Geological Survey (USGS) maps are projected using some version of UTM.

3.4.4.3 State Plane Coordinate System (SPCS)

Although the general level of distortion for UTM is low, for many applications, minimizing distortion and maximizing accuracy is of paramount importance. Driven by this need, the SPCS provides minimally distorted projections that are frequently used for surveying applications, utility placement, and the like. Each state in the U.S. had its coordinate system and based on the state's shape, the appropriate projection was chosen. For example, the state of Texas is so large, it consists of five state plane zones (North American Datum [NAD] 1983) that are based on the Lambert conformal conic projection (Texas Parks and Wildlife Department [TPWD], 2011). Conversely, the state of Arizona consists of three zones (NAD, 1983), all of which use a transverse Mercator projection (Fig. 3.11). SPCS are the best options for local distance calculations and spatial analysis. Versions that use both the metric and imperial systems of measurement are available for mapping.

3.4.4.4 Lambert Conformal Conic

One last projection worth detailing is the Lambert conformal conic. For the coterminous United States, the Lambert conformal conic uses the standard parallels of 33°N and 45°N. This keeps the actual scale distortion relatively small, even for the map edges (~3%) (Kimmerling et al., 2011). This projection is best for countries and states that display a decidedly east-west trend in their geographic expanse. For

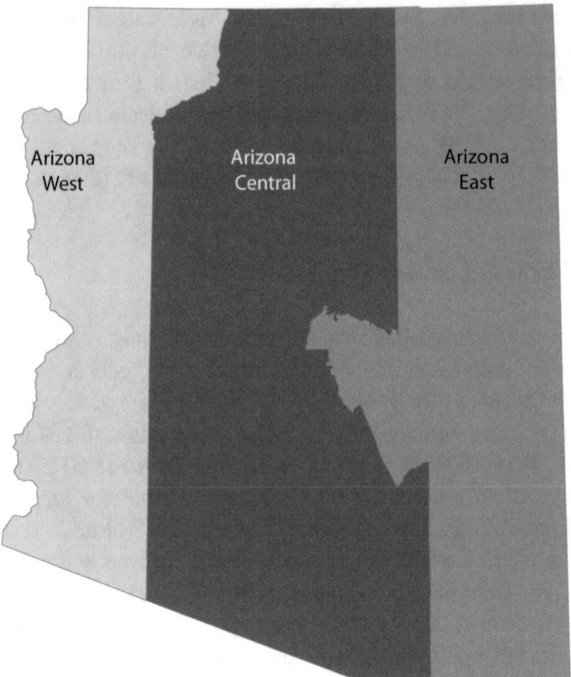

Fig. 3.11 State plane coordinate system for Arizona (Projected as Central)

example, the state of Oregon uses a SPCS based on the Lambert conformal conic projection. The other major application of this projection is for navigational purposes. At the 1:500,000 scale, a straight line drawn on a map using the Lambert conformal conic projection is almost identical to a great circle. As the name suggests, this projection preserves shapes and directions locally.

3.5 Georeferencing

Given the sheer scope of material covered thus far in this chapter, it would be easy for readers to lose track of all the important information regarding representation, geodesy, and projections. To be sure, the material is multifaceted, dense, and technically challenging. But as promised, we have endeavored to make this material as accessible as possible, particularly for readers who do not have any plans to pursue an advanced degree in Cartography or Geodetic Engineering. In an effort to tie all of these concepts together, we have saved the topic of georeferencing for last. Georeferencing refers to the process of assigning a location to atoms of GI (Longley et al., 2005). Again, this may consist of a latitude and longitude for representing the location of a burglary, or it may simply consist of a specific and unique placename.

Regardless of its form, it should be abundantly clear to readers that this process is deceivingly complex, requiring elements of spatial representation, geodesy, and map projections to do it accurately, and without distortion or bias.

As one might expect, a range of georeferencing systems exists, from placenames to cadasters and the Public Land Survey System (PLSS) for the U.S. All of these systems have strengths and weaknesses, but as mentioned previously, they also share a common undercurrent in that they are designed to help differentiate and reference specific locations. Given space constraints, we will not endeavor to address all of the georeferencing systems widely deployed in the U.S. or elsewhere. However, we will highlight two of the most popular, placenames and postal addresses and codes, stressing their strengths and weaknesses. Further, although this review is U.S.-centric, most countries have a similar array of georeferencing systems for the built environment.

As detailed by Longley et al. (2005), there are really only three major requirements for georeferencing: (1) the locational data need to be unique, helping to avoid confusion and eliminate ambiguity; (2) the georeference must also be recognized and its meaning shared amongst all people and agencies that wish to use the information; (3) the georeference must be persistent through time, thereby eliminating the need to update this information across multiple, independent agencies (or people) that rely upon the georeferences for operation.

3.5.1 Placenames

The use of placenames is one of the most straightforward ways to georeference locations. A placename is a word or series of words that identify physical or administrative features on the Earth, sea floors, or other planets (Randall, 2001). Some placenames are broadly recognized, but many others are not. Regardless, their use provides a powerful tool for establishing georeferences at the local, regional, and global scales. For example, consider a relatively local georeference in Phoenix, Arizona—Murphy's Bridal Path. It is a crushed rock trail that runs along the east side of Central Avenue through North Central Phoenix. The vast majority of residents in this section of Phoenix know where this trail is located and how it is used, but for people living in San Antonio, Texas, Vancouver, Canada, or even Mesa, Arizona, it is not a particularly meaningful georeference. Place naming at local levels are endonyms, terms used by people who live there and in their language. Conversely, there is no doubt that the Pacific Ocean is a globally recognized georeference. However, the placename itself varies by language (e.g., Oceano Pacifico [Spanish]; Stillehavet [Danish]; Pacifik [Croatian]; and so forth). This is one of the major limiting features of placenames—although the georeference is consistent, the actual placename changes given the culture and/or language describing the same place. In some extreme cases, some cultures and languages may not have specific enough terms for geographic features of new places. For example, the Yindjibarndi people of the Australian outback have terms for dry creek bed (i.e., wundu) and concavities in hillsides (i.e., garga) because during extreme weather, water accumulates in those places and water is scarce in the outback (Mark, Turk, & Stea, 2007).

For georeferencing, placenames are key because people think in terms of them (Goodchild, 2011). The standardization of placenames in the U.S. began in 1890 when the U.S. Board on Geographic Names (USBGN) was created to keep placenames accurate, consistent, and disambiguate multiple local terms for the same place. A supranational group also exists, the United Nations Group of Experts on Geographical Names (UNGEGN), to lend authority control to global naming. The goal of consistent naming (i.e., each named feature has a single authorized name) are to discourage the duplication of placenames and allow for coordinated exploration, mining, settlement, homeland defense, emergency response, and taxation (Bishop, Moulaison, & Burwell, 2015). During the formal documentation, the georeferencing includes assigning coordinates to a placename (Buchel & Hill, 2011).

Another limitation with using placenames as georeferences is associated with their scale and scope. For example, if one uses the placename "Midwest" in the United States, it is not immediately clear which part of the country that one is referencing. As a result, a fundamental question concerning scope emerges; which states are included in the Midwest? Kurtzleben (2014) provides 41 different graphics that help explain how the Middle West is defined and which states it might include (Hint: Illinois, Indiana, and Iowa are solidly Midwestern). However, even if there was a firm definition regarding the geographic scope of the Midwest, its use as a georeference is still lacking. For the sake of argument, let us include Illinois, Iowa, and Indiana only. These three states consist of 150,604 square miles when combined. Thus, given the sheer scale of this version of the "Midwest", a georeference that uses this placename does not provide much specificity regarding location.

One last limitation of placenames is their variation in temporal persistence. For example, in 1916, the city of Hot Springs, New Mexico was incorporated in Sierra County. However, in March of 1950, Hot Springs changed its name to Truth or Consequences to win the opportunity to host an episode of the once popular gameshow of the same name. Placename changes, however, are not limited to small towns in New Mexico. To demonstrate power over places and claim governance, placename renaming occurs throughout history and will not likely stop. For a country specific example, the Ukraine changed 90% of its place-names back to original Ukrainian names when the country became independent in 1991. For now, it is unclear how many of those retro-placenames for cities, bodies of water, and transportation, will remain or revert back to Soviet-era names. There are other items to complicate things, such as few countries have one singular language, mistranslations occurring in Romanization of placenames, and complications with writing systems beyond alphabets. For example, the Chinese capital, Peking, was changed to Beijing in 1958. Interestingly, there was no formal placename change in China as the character remains the same. After the establishment of the People's Republic of China in 1949, the government adopted the pinyin transliteration method for writing proper names (e.g., placenames, people's names, and so forth) and this used the Latin alphabet. So, while Peking was transformed to Beijing in the English language, in Chinese, the name remains exactly the same.

3.5.2 Postal Addresses

The georeferencing system for postal addresses is both simple and efficient, leveraging hierarchies of information for its functionality. In short, postal addresses draw upon a combination of placenames and more formal numbering systems that subdivide regions and their places into smaller, more manageable bits. Longley et al. (2005, p. 113) provide a good summary of this, detailing the basic assumptions:

1. Every residential dwelling and commercial entity with physical space is a potential destination for mail;
2. These dwellings and offices are arrayed along paths, roads, streets, or waterways (Roney, 2013) and are numbered accordingly;
3. The names of paths, roads, streets, and waterways are unique within local areas. Prefixes (e.g., E, W, N, S) and suffixes (St., Dr., Pkwy, and so forth) help with this;
4. The local areas (e.g., cities and towns) have names that are unique within larger regions;
5. Regions have names that are unique within countries (e.g., states).

One of the more overlooked, but critically important components of the postal address system are Zone Improvement Plan (ZIP) codes. The most common form of the ZIP code, which uses five digits, was implemented in the early 1960s in the U.S. to facilitate a more efficient routing of mail and secondarily a georeferencing system for addresses throughout the country. There are a number of very interesting quirks to the ZIP code system, including the use of 3-digit ZIPs that correspond to regional sorting centers (i.e., sectional center facilities) that serve multiple states, to the ZIP+4 codes, which add four additional digits (bringing the total to 9) that often correspond to specific blocks or high-volume mail receivers (e.g., corporate or university campus) within the 5-digit ZIP service zone. For more details on ZIP codes, their history, and how they can and/or cannot be used effectively for geographic analysis, see Grubesic and Matisziw (2006) and Grubesic (2008).

Consider for example a typical postal georeference for the United States: 5 Dartmouth Circle, Swarthmore, PA. Does this georeference meet the three basic requirements set forth by Longley et al. (2005) for georeferencing? More importantly, does it leverage the power of information hierarchies for georeferencing? The house number and street pairing "5 Dartmouth Circle" is not unique. For example, this pairing exists in at least six other locations throughout the U.S., including Longmont, CO 80503, Pembroke, MA, 02359, Odessa, TX 79764, Bedford, NH 03110, Oak Ridge, TN 37830 and Media, PA 19063. However, when one adds the city (Swarthmore), state (PA), and ZIP code (19081), the georeference improves dramatically. Indeed, with that additional information, one can easily geolocate the house, and recognize that there is no other house/structure at that address. *It is unique*. That said, there is another georeference in the region that is nearly an exact match—5 Dartmouth Circle, Media, PA 19063. These locations are less than four miles apart in suburban Philadelphia, and their similarity should emphasize how important it is to have an explicit suite of georeferencing conventions that are, in fact unique. That said, these two addresses, although quite similar, leverage the power of information hierarchies and can be differentiated by city and ZIP code to ensure that there is no confusion between the georeferences.

Is the meaning of this georeference shared? For all of the U.S., the meaning of a postal address is shared. Mail is easily delivered to 5 Dartmouth Circle, Swarthmore, PA 19081. In addition, based on this georeference, taxes are levied, elementary, middle, and high schools are assigned, and navigation to this location is possible. However, as detailed by Longley et al. (2005), the georeference is less meaningful for people living in places like China where the Roman alphabet is not used. For example, consider the following address from China (in Mandarin) (Grigg, 2011):

中国,山东省,青岛市 香港东路6号,5号楼,8号室 李小方 (先生)收

The address convention in China is reversed from that in the United States. In this instance, one starts with symbols that represent the "People's Republic of China" or "PRC", then moves to the province of Shandong, then the city of Qingdao, and so on. Perhaps most disconcerting to Westerners is the Chinese symbology used for this georeference. For those who are unfamiliar with the language, the symbols detailed above could mean anything. The same could be said for residents of China when they see a georeference using the Roman alphabet. For these reasons, the georeference is not *widely* shared, but agencies that specialize in the trading of such information can interpret the associated georeferences and make their best attempt to, as an example, deliver post to each location.

Lastly, is the Swarthmore georeference persistent through time? According to the auditor records for Delaware County, PA (2016), the house at 5 Dartmouth Circle was built in 1952 and the georeference has not changed since it was assigned nearly 65 years ago. For the U.S., this qualifies as persistence. However, less than five miles away from the Swarthmore address is the historic Lower Swedish Cabin (Halligan, 1989) located on Creek Rd, Drexel Hill, PA 19026. Built in 1654, the formal georeference for the cabin has changed several times over the centuries as municipalities were formed, and streets were built in the region. To provide a bit of temporal context here, the Lower Swedish Cabin was built during the same decade that Rembrandt was actively painting, and both the Russo-Polish War and Anglo-Spanish War began. Thus, it is not surprising that the georeference has changed on occasion.

As noted above, there are many other georeferencing systems in use, both in the U.S. and elsewhere. Some are globally unique, while others are only relevant for a local area. The same can be said for their scale and scope, because the spatial resolution for all of these systems can vary widely. Beyond the built environment, georeferencing becomes fuzzier, but tools do exist. For example, GEOLocate provides biocollections (i.e., ~three billion specimens globally and counting) the means to identify point localities, estimate uncertainty radii, and delineate likely habitat polygons (e.g., lake) (http://www.museum.tulane.edu/geolocate/default.html). The tool works by pulling placenames from locality metadata, then detecting water bodies, bridge river crossing features, and direction from these known locations in conjunction with base maps from several common Geospatial Web (Geoweb) tools to georeference biota (e.g., Tussahaw Creek at LeGuin Mill Road, approximately 3.5 mi.

ENE Locust Grove - Segment 3., USA, Georgia, Henry ≈ 33.383316, −84.032236). Regardless of the system examined, the three core tenants of georeferencing remain important and critical to operation: (1) uniqueness; (2) recognized, with a shared meaning; and (3) temporal persistence.

3.6 Conclusion

Given the wide range of material covered in this chapter, perhaps it is apparent that geographic representation and the science of geodesy is critical for developing, managing, organizing, consuming, analyzing, and visualizing GI. From projection systems and geometry to the graticule and georeferencing, all of these concepts and tools are fundamental, complimentary, and interrelated. It is worth noting that all of these formalities, rules, and caveats in using GI, while helpful, were not always necessary or needed. For example, at one point in time, the Chamoru people of the Mariana Islands used four different directions. There was seaward (lagu), inland (haya), right of seaward (kattan) and left of seaward (luchan). There were no fixed points and no use of a compass; the only references were toward the sea or toward the land. There was no need for more directions because the entire world for the Chamoru consisted of their island. To be sure, after exploring the material in this chapter many readers will wish to be teleported back to a simpler time where only four directions were needed for all geographic representation. However, recent advances in GI data collection and the drive to better understand our world means that more complex systems are required to measure and compare the earth and its features. Lastly, we realize that many users will not have all of the required technical information in hand when exploring their GI, but this chapter is intended to serve as the jumping off point for tracking down the key information and tools required to dig a bit deeper analytically and represent the results more effectively.

References

Bishop, B. W., Moulaison, H. L., & Burwell, C. L. (2015). Geographic knowledge organization: Critical cataloging and place-names in the Geoweb. *Knowledge Organization, 42*(4), 199–210.

Brunsdon, C., & Corcoran, J. (2006). Using circular statistics to analyse time patterns in crime incidence. *Computers, Environment and Urban Systems, 30*(3), 300–319. doi:10.1016/j.compenvurbsys.2005.11.001.

Buchel, O., & Hill, L. L. (2011). Treatment of georeferencing in knowledge organization systems: North American contributions to integrated georeferencing. *Knowledge Organization, 37*(1), 72–78.

Bugayevskiy, L. M., & Snyder, J. P. (1995). *Map projections, a reference manual*. London: Taylor & Francis.

Chang, K. T. (2006). *Introduction to geographic information systems (pp. 117–122)*. Boston: McGraw-Hill Higher Education.

Chrisman, N. R. (1997). *Exploring geographic information systems*. New York: Wiley.

Delaware County, Pennsylvania. (2016). *Auditor records*. Retrieved from http://tinyurl.com/zn6xqw4

Fisher, N. I. (1995). *Statistical analysis of circular data*. Cambridge: Cambridge University Press.

Fisher, P. (1997). The pixel: A snare and a delusion. *International Journal of Remote Sensing, 18*(3), 679–685. doi:10.1080/014311697219015.

Goodchild, M. F. (2011). Formalizing place in geographic information systems. In L. Burton, S. Kemp, M.-C. Leung, S. Matthews, & D. Takeuchi (Eds.), *Communities, neighborhoods, and health* (pp. 21–33). New York: Springer.

Grigg, H. (2011). How to write a postal address in Chinese. *East Asia student: Random stuff related to East Asia*. Retrieved from http://tinyurl.com/zaskgwh.

Grubesic, T. H. (2008). Zip codes and spatial analysis: Problems and prospects. *Socio-Economic Planning Sciences, 42*(2), 129–149.

Grubesic, T. H., & Matisziw, T. C. (2006). On the use of ZIP codes and ZIP code tabulation areas (ZCTAs) for the spatial analysis of epidemiological data. *International Journal of Health Geographics, 5*(1), 58.

Halligan, T. (1989). A 335-year-old Drexel hill cabin gets its due: A place in history. *Philly*. Retrieved from http://articles.philly.com/1989-10-26/news/26117988_1_trading-post-cabin-site-swedes.

Harary, F. (1969). *Graph theory*. Reading, MA: Addison-Wesley.

Howse, D. (1980). *Greenwich time and the discovery of the longitude*. New York: Oxford University Press.

Kimmerling, A. J., Buckley, A. R., Muehrcke, P. C., & Muehrcke, J. O. (2011). *Map use: Reading, analysis, interpretation* (7th ed.). Redlands, CA: Esri Press Academic.

Krygier, J., & Wood, D. (2011). *Making maps: A visual guide to map design for GIS*. New York: Guilford Publications.

Kurtzleben, D. (2014). 41 maps and charts that explain the Midwest. *Vox*. Retrieved from http://tinyurl.com/q7z6zgv.

Longley, P. A., Goodchild, M. F., Maguire, D. J., & Rhind, D. W. (2005). *Geographic information science and systems*. Chichester, England: Wiley.

Mark, D. M., Turk, A. G., & Stea, D. (2007). Progress on Yindjibarndi Ethnophysiography. In S. Winter, M. Duckham, L. Kulik, & B. Kuipers (Eds.), *Spatial information theory: 8th International Conference, COSIT 2007, Melbourne, Australiia, September 19–23, 2007. Proceedings*. (pp. 1–19). Berlin: Springer.

McDonnell, R., & Kemp, K. K. (1995). *International GIS dictionary*. Cambridge, England: GeoInformation International.

Murray, A. T., & Grubesic, T. H. (2012). Spatial optimization and geographic uncertainty: Implications for sex offender management strategies. In *Community-based operations research* (pp. 121–142). New York: Springer.

Olsen, N., Lühr, H., Sabaka, T. J., Mandea, M., Rother, M., Tøffner-Clausen, L., et al. (2006). CHAOS—A model of the Earth's magnetic field derived from CHAMP, Ørsted, and SAC-C magnetic satellite data. *Geophysical Journal International, 166*(1), 67–75. doi:10.1111/j.1365-246X.2006.02959.x.

Olsen, N., & Mandea, M. (2007). Will the magnetic north pole move to Siberia? *Eos, 88*(29), 293–294.

O'Sullivan, D., & Unwin, D. J. (2010). *Geographic information analysis* (2nd ed.). Hoboken, NJ: Wiley.

Randall, R. (2001). *Place names: How they define the world—And more*. Lanham, MD: Scarecrow.

Robinson, A. H., Morrison, J. L., Muehrcke, P. C., Kimmerling, A. J., & Guptill, S. C. (1995). *Elements of cartography* (6th ed.). New York: Wiley.

Roney, M. (2013). In Alabama town, snail mail comes by boat. *USA Today*. Retrieved from http://tinyurl.com/gvwapfc.

Texas Parks and Wildlife Department [TPWD]. (2011). *State of Texas State Plane Zones*. Retrieved from http://tinyurl.com/jl578xh

Utrilla, P., Mazo, C., Sopena, M. C., Martínez-Bea, M., & Domingo, R. (2009). A palaeolithic map from 13,660 calBP: Engraved stone blocks from the Late Magdalenian in Abauntz Cave (Navarra, Spain). *Journal of Human Evolution, 57*(2), 99–111.

Chapter 4
Policy

Abstract The purpose of this chapter is to provide an overview of geographic information (GI) policy. A review of the Spatial Data Infrastructure (SDI) concept gives readers a framework for government GI organization, access, and use policy issues. The chapter concludes with an in-depth analyses of the U.S. National Spatial Data Infrastructure (NSDI), which summarizes one country's past approaches and future directions for SDI implementation.

4.1 Introduction

Information policy influences the entirety of geographic information (GI) organization, access, and use, and critically shapes both humanity and its interactions with most forms of information. Because information is increasingly mediated by networks (e.g., the Internet), including many day-to-day functions and community activities, the policies dictating how to create, handle, and share information impacts most (if not all) people in the developed world. An important facet of information policy is to direct the progress and corral the extent of information uses. This includes policies for the mundane, including both the creation and disposal of government documents. Consider, for example, the wide range of U.S. federal policy generated on a daily, weekly, monthly and yearly basis, from formal legislation, "executive orders, judicial rulings, guidelines and regulations, to rulemaking, agency memos, signing statements, agency circulars, and other types of official statements" (McClure & Jaeger, 2008, p. 257). Topically, these policies could relate to broad issues such as pricing, security, and privacy, as well as finer points of information management that appear in a data lifecycle: (1) creating or receiving data; (2) appraising and selecting data; (3) ingesting; (4) preserving; (5) storing; (6) accessing, using, and re-using data; and (7) transforming data (Higgins, 2008). As the quantity of open government data continues to grow, information management faces new challenges. In particular, as the potential users of open government data attempting to discover, identify, select, and obtain information, it will be more diverse, varied, and further removed from the creators. This is especially true for GI.

This introduction presents an overview of information policy and the associated concepts such as stakeholders, open government, and others, in the context of GI. A U.S. national scale information policy discussion gives one example of pol-

W. Bishop, T.H. Grubesic, *Geographic Information*, Springer Geography,
DOI 10.1007/978-3-319-22789-4_4

icy analysis with GI management issues. That examination sets the stage for later sections reviewing the broader GI policy concept of the Spatial Data Infrastructure (SDI), its value, costs, users, and communities. A SDI is the coordination of GI creation, collection, dissemination, and storage *between* stakeholders, in a larger community (Williamson, Rajabifard, & Binns, 2006). The success of SDI coordination is vital to any GI user or organization managing GI because all potential users and organizations: (1) need more GI than they can afford to create or collect on their own; (2) need GI outside their jurisdictions or operational areas; and (3) need GI to be compatible for use. The chapter concludes with the background, policy analysis, and future considerations for SDI implementation in the U.S.—the National Spatial Data Infrastructure (NSDI)—the first national SDI in the world.

4.1.1 Information Policy

One of the unique features of information policy is its tendency to crosscut other forms of policy, including environmental, transportation, telecommunication, and the like. As a result, no central body of law or administrative unit in a government coordinates or oversees information policy, let alone GI policy. However, GI policy is unique in that it does rely on a handful of federal agencies, those with GI central to their missions, to craft and implement forms of GI policy in the U.S. (e.g., USGS). Like all policy, there are various stakeholders involved. Some have conflicting viewpoints on how to address potential problems with GI policy—that is, if they can agree whether or not a problem exists in the first place. In this context, *status quo* may benefit some stakeholders because any policy change jeopardizes their present stakes and/or agendas.

Not surprisingly, the complexity of policy and its associated implications increase with the number of stakeholders. This includes any and all agencies involved with any aspect of GI creation and management. In addition, as the number of users seeking access to developed GI increases, the gravity of decisions regarding GI policy becomes more significant. For example, policy changes for American Community Survey (ACS) data would impact a much larger audience of users than would changes to the Conservation of Arctic Flora and Fauna (CAFF) data.

Information policy would not require an entire chapter if all government information were open or closed. At present, these open and closed extremes do not exist in any reality. For proprietary GI, or that related to classified geospatial data activities pursued by the Department of Defense, the discussion is short. A publisher or provider simply owns the GI and may choose to share it, or not. For example, some proprietary data, such as consumer spending patterns collected from national geodemographic surveys, are available for purchase from secondary data vendors. However, for non-proprietary and non-classified GI, the conversation becomes more nuanced and complex. For example, in the U.S., federal GI is produced directly or indirectly using tax dollars. However, not all of the GI produced with tax dollars is open and available. There are many reasons driving data openness (or not) and the particulars of government information ownership and its controversies can be found in other sources (Evans, 2011; Priest & Arkin, 2010; Sunstein, 1986). Regardless of the policy decisions driving open/closed data decisions, it is important to consider the context of such decisions. For example, in some cases where the GI is open and available, the distribution platforms do not receive the same level of usability testing or promotion as their commercial counterparts. This results in decreased discoverability, access, and use in many cases. In other instances, platforms are built using commercial tools but without the resources to upgrade when a sea change in technology occurs (Bishop, Haggerty, & Richardson, 2015). As a result, even though the data are "open", their distribution platform is antiquated and less accessible. Sadly, these types of scenarios occur frequently with many types of information. Power users experience lock-in with archaic systems and do not see a need to update systems. This mentality may inhibit any expansion of access and creative uses of GI beyond their original system functions for select audiences.

To be sure, the degree of openness or closedness of government information are central to information policies in the U.S. These policies are also frequently contradictory. For example, parts of the USA PATRIOT Act of 2001 (2001) (P.L. 107–56) and the longstanding Privacy Act of 1974 (1974) (P.L. 93–579) are at odds over *what* and *how* personal information can be collected and shared within the federal government and with individual citizens. Big data further complicates privacy as more people consensually share participatory personal information with social media providers (e.g., Facebook) to gain social capital (Crampton, 2015; Ellison, Steinfeld, & Lampe, 2007). Of course, consumers are still free to *not* share their personal information, especially since much of it is made available to the U.S. intelligence community upon request. In the end, all policies represent some form of compromise between stakeholders with opposing perspectives and the policymakers themselves. The personal privacy versus national security conundrum has dogged U.S. information policy since 2001, but it is not unique to any one country. These tradeoffs have been evaluated by many federal governments and the resulting policy frameworks that emerge from these debates are, at least to some extent, distinctive for each nation. In turn, policy researchers can evaluate the efficacy of information policy by assessing the implementation, enforcement, impact, and emergent gaps after policy creation. The findings of these analyses can be used to inform new or revised policies that help address core problems as their circumstances and context evolve.

We will revisit these topics toward the end of the chapter, summarizing U.S.'s historical approach to government GI organization, access, and use, and providing a detailed policy analysis of the last thirty years stemming from the implementation of the NSDI. However, we first present a brief policy analysis of one broadly generalizable problem that touches upon many of the most important social, economic, and technological issues of our time and includes a considerable amount of GI—broadband and the digital divide (Mossberger, Tolbert, & McNeal, 2008; Prieger & Hu, 2008).

4.1.2 The U.S. Broadband Policy Example

Compared to the multifaceted policy issues associated with privacy and security, where finding the middle ground is difficult, U.S. broadband policy provides an instance where most stakeholders and citizens understand that overcoming the digital divide will require elements of government strength to mitigate. Further, there are more direct channels for evaluating the relative success and/or failures of broadband policy than other policy implementations. As mentioned earlier, for the purpose of this chapter, broadband policy also provides a manageable example to introduce some basic concepts pertaining to the NSDI without getting buried in the enormity of the guidelines and regulations that dictate organization for the corpus of national GI.

To oversimplify, the goal of U.S. broadband policy is to connect all citizens to the Internet, with sufficient speeds to download and upload data, voice, video, and other information. The costs must also be reasonable because equal participation is key to a representative democracy in the twenty-first Century. The route taken to arrive at this goal relates to both national pride and the *triple bottom line*, which are both advantages for its successful implementation. The triple bottom line is an accounting framework that considers the *economic*, *social*, and *environmental* drivers of performance (Elkington, 1998). The reason it is effective for evaluating information policy is that most stakeholders and governments are typically invested in one of the three bottom lines (Rajabifard & Williamson, 2001). In the context of telecommunications, when allowing market forces to determine the rollout of broadband networks, many rural and remote communities, particularly those with thin (i.e., low population) markets, were left with limited (or no) service options, especially when compared to thicker, more profitable urban markets (Grubesic & Mack, 2015).

For example, early in the process, among households with Internet access, rural penetration rates of 7.3% were far short of those in the central city (12.2%) and urban (11.8%) locales (National Communications and Information Administration (NTIA), 2000). Today, access for the most rural portions of the U.S. remain a concern, but more nuanced problems have emerged. Namely, rural and remote locales continue to use technological platforms that are at least one or two generations behind their urban counterparts (Grubesic & Mack, 2015). For example, although fiber-to-the-curb (FTTC) is common in many metropolitan areas, it has been slow to arrive in rural America (Grubesic, 2016). The overarching concern here is that communities left without first-tier access will suffer from economic decline and will not be able to participate equally in an e-government centered democracy that determines social and environmental futures of communities (Mossberger et al., 2008). Again, this trend persists as technological advancements outpace deployment (Chester, 2007). Elements of national pride also assisted in drawing recognition to the challenges inherent to broadband deployment in the U.S. For example, in 2010 the broadband penetration rate for the U.S. was 27.1 per 100 inhabitants, ranking 14th globally (OECD, 2010). This was not surprising to those involved in the broadband industry, but it did serve as a wakeup call to the general public.

In an effort to mount a national response lagging broadband penetration and infrastructure provision in the U.S., the Federal Communications Commission (FCC) released the National Broadband Plan (NBP) as a framework to help "stimulate economic growth, spur job creation, and boost America's capabilities in education, healthcare, homeland security and more" (Federal Communications Commission [FCC], 2010). Like all triple bottom line efforts, NBP sought to promote the construction of infrastructure that facilitates improvements in "health care, education, energy use, economic opportunity, government performance, civic engagement and public safety" (Grubesic, 2012a, p. 114). The NBP's call to action was complex and included the development of a national broadband map, which was to serve as a comprehensive inventory of GI for broadband infrastructure, providers and other quality of service metrics (e.g., speed) at the block level for the entire U.S. (http://www.broadbandmap.gov/). In the end, although the overall quality of the national broadband map (NBM) data were questionable (Grubesic, 2012a, 2012b), the motivation to base broadband policies on empirical information was solid. The data were made available to both interested researchers and the general public through a basic map/geovisualization interface (see above) and for tabular download. The tabular data were organized by individual states and broadband deployment data were summarized at the Census block level. Although space constraints prevent us from detailing all the problems associated with these data, they suffered from significant spatial uncertainty when delineating the geographic coverage of broadband, used mixed Census geographic base files (e.g., block tessellations from both 2000 and 2010) and failed to use standardized geoprocessing rules for determining broadband service constraints (Grubesic, 2012a, 2012b). In this example, we recognize that the data quality, accessibility, and usability of the broadband map, and long-term preservation of the GI, were all secondary concerns to the overarching policy and its goals. However, even with all of this formal effort from a large government agency, interested stakeholders, citizens and the private sector, as well as having a clear and measurable goal of broadband access for all U.S. citizens, this example clearly shows how difficult it can be to conduct a systematic evaluation of information policy and its motivation at a national scale as well as the implications to its associated GI. There are many moving parts, problematic data, and real technological hurdles to consider. All of which complicate the systematic evaluation of one well-defined information policy at a national scale. Therefore, the policy analysis of the NSDI cannot possibly detail the vast array of GI impacted by that policy or all the potential stakeholders at the granularity conducted here for broadband.

4.2 Spatial Data Infrastructure (SDI)

The term SDI can be applied to any *de facto* spatial data plan for nations, organization, or institutions. Again, SDI does not relate to any one data warehouse or database but is the result of harmonization in the creation, collection, dissemination, and storage of GI *between* stakeholders, in a broader geospatial data community

(Williamson et al., 2006). SDIs are complex and it is difficult to observe them without considerable abstraction. Frankly, it is even difficult to create a flowchart that fits neatly on one screen or page given the number of moving parts for most SDIs. To complicate matters, when writing about SDIs, most authors take relatively varied approaches to address core concepts. The inclusion or exclusion of a combination of data-related objectives, broader objectives, and/or components in definitions make creating a single accepted definition difficult (Hendriks, Dessers, & Van Hootegem, 2012). To clarify, the term SDI does not typically apply to something as small as one geoportal, but the term has been ascribed to all kinds of geospatial data collections. In effort to simplify the following SDI discussion, we will focus on SDIs that organize government GI. This is a reasonable subset of SDIs because most governments face similar challenges in establishing and operating this type of infrastructure.

Government SDIs are a complex concept because: (1) the implied hierarchies from local to federal government and all combinations of relationships beyond the government; (2) ill-defined users and communities; (3) functional and adoption objectives; and (4) its many components (e.g., technology, systems, standards, networks, people, policies, and so forth) (Hendriks et al., 2012). The ingredients for each SDI come from the mindboggling number of socio-technical components that relate to any GI involved in government operations. These operations often overlap in coverage, compete for resources, and have disparate goals (e.g., conservation and development).

The central purpose for any SDI is to facilitate and coordinate the exchange, sharing, accessibility, and use of GI across stakeholders (Theraulaz, Gautrais, Camazine, & Deneubourg, 2003). Data-handling facilities, and any entity interacting with the institutional, organizational, technological, human, and economic resources, generally fall under the purview of an SDI (Erik de Man, 2006). The technological infrastructure and its components facilitate the functional objectives of making GI organized and accessible, but nearly all SDI definitions combine broader objectives and components beyond the technology to inform their design and development. However, because technology continually evolves, dialog about technology within the domain of SDI gets technical, very quickly. Since rudimentary functionality is required for any SDI to work, we dedicate our efforts in this chapter focusing on the broader concepts inherent to the value of SDI and its users.

The broader objectives that drive SDI creation stem from the primary purpose of all national and global SDIs—*to disseminate GI between stakeholders and across communities in an accessible and useful way.* Rajabifard and Williamson (2001) succinctly present core problems that a successful SDIs should solve:

- Most organizations need more GI than they can afford;
- Organizations often need GI outside their jurisdictions or operational areas; and
- GI collected by different organizations are often incompatible.

On the face of it, sharing GI sounds like a simple task—have data, will share. Such utilitarian sentimentalities detract serious critique. In practice, any SDI creation requires agreements on (1) what GI should be shared; (2) how that GI will be shared; and (3) who pays for which infrastructure that enables sharing. Put succinctly and without complex definitions, the design and development of SDIs should focus on "real people using real systems to address real problems" (Carver, 2003, p. 68). In fact, for any GI sharing to occur, geospatial data needs to be created in compatible formats, types, or transformable for use by the entities within the same infrastructure.

Deciding *what* and *how* to share GI should arise from the types of real problems that real users need to address. The processing platforms for GI, including geographic information systems (GIS) and related statistical analysis packages are now less of a concern because most of them have a strong suite of basic analytical and visualization tools. Therefore, instead of worrying about storage, processing, and analysis, many users are more concerned with the acquisition of quality GI (Rhind, 1999). As reviewed in Chaps. 2 and 3, certain types of GI are difficult and costly to collect. With increased Internet connectivity, bandwidth and more flexible formats, however, all GI has become much easier to share.

Considering that a primary function of governments is to make decisions related to complex problems, information systems are absolutely essential to the process. That said, many real problems are inherently geographic and require GI sharing in information systems, federal or otherwise. The GI shared through SDIs is a subset of the GI that has been collected by government entities. For many nations, a systematic aggregation of local GI occurs to address broader national issues. Obviously, each nation has a different history and unique local context for the needs and subsequent development of SDI. But, in general, most countries present a centralized structure with certain ministries, agencies, bureaus, and any number of bureaucratically named structures charged with the collection of certain government GI. Regardless of structure, all governance propagates similar GI collection that roughly relate to a delineation of land ownership, quantifying of natural resources, and description of humans and structures in the built environment.

Government GI also has a long history of being repurposed into cartographic resources to share with decision-makers. However, sharing GI beyond decision-makers and static print maps has introduced complications. Geographic information creators, allied geographic information professionals, and decision-makers that now have access to GI may independently use these resources to generate their own analyses, visualizations, and conclusions (i.e., broadband deployment). Within this context, a democratization of government GI becomes a reality, especially as more potential users have access to GI (Dunn, 2007). In fact, many non-professional users may not have the authority or knowledge to employ GI for other purposes. These users may introduce representations and worldviews that differ from the government and its purposes. As a result, governments strategically limit access to SDIs to avoid uncomfortable or conflicting representations of land as well as security and privacy concerns.

One way to limit uncertainty and encourage a controlled message when using GI is when governments choose to only share between governmental agencies. This certainly helps avoid the pitfalls related to the democratization of GI, but many technical problems remain, including (1) standard metadata; (2) standardized semantics (i.e., use of the same terminology); (3) automatic updates; and (4) exchange (i.e., real-time delivery) (Peng, 2005). Advancements in technology do not alleviate institutional barriers to GI sharing (e.g., information hoarding)

(Obermeyer & Pinto, 2008). Once institutional and interoperability issues are managed, challenges associated with usability and findability of GI for expert and non-expert users remain.

Consider, for example, recent experiences regarding the lack of GI sharing across governments during disaster response efforts. Large-scale disasters (e.g., terrorist attacks, earthquakes, floods, bridge collapses or hurricanes) require the coordinated efforts of a variety of organizations (e.g., federal and state agencies, volunteer organizations, and residents) to respond quickly and minimize loss (Jaeger et al., 2007). To illustrate, "1,607 governmental and non-governmental organizations responded to the 9/11 terror attacks in New York City," but incompatible technologies (e.g., radio systems) did not allow individuals from different organizations to communicate with each other (Kapucu, 2004). A few years later, the inability of professional emergency responders (PER) to communicate left thousands of Katrina survivors stranded in the New Orleans Convention Center until the Federal Emergency Management Agency (FEMA) was notified by the news media (Jaeger et al., 2007). These failures resulted, in part, from a lack of real-time GI sharing between U.S. government decision-makers. In addition, more long-term government purposes also could benefit from sharing GI through SDIs. Clearly, when disaster strikes, both the economic and social cost of not having a working SDI is clear, but the valuation of government GI for everything else necessitates more exploration.

4.2.1 Value

The monetary costs of creating and managing GI are significant, but the costs associated with not sharing these data, in a fluid manner, significantly elevate the base price. For example, during disasters, the GI collected for other purposes, such as water management and flood prevention mitigation, can be critical pieces of the first responder and disaster mitigation puzzle. Again, the triple bottom line is an important construct here. It is a largely unmeasured value justification that drove the creation of many SDIs, but has never been truly evaluated or tested using rigorous empirics. However, the U.S. the Office of Management and Budget (OMB) recently called for research to formally measure the return on investment (ROI) of the NSDI. To be sure, all SDIs would benefit from a better understanding of value and cost of sharing GI. Considerable work has been accomplished by SDIs assuming that data silos were barriers to advancements in a number of areas. But, as geoportals are developed or dissolved, a greater understanding of these pieces of the cyberinfrastructure is required to justify the related creation and maintenance costs of SDIs, as well as the intrinsic values of sharing GI.

An alternative way to think about the value of GI is to consider its development in the context of markets. The users and associated communities that need GI for different purposes form these markets. However, the markets do not function like those detailed in the neoclassical economics literature. For one, fewer users and smaller communities require global geospatial data, reducing both supply and demand, as well as market size. Second, the type of problems that routinely require

GI tend to occur at local or regional scales, but the GI is often coordinated at higher levels of governance. For example, during disasters, emergency managers and their organizations (e.g., FEMA) need access to significantly more GI than an individual waiting for the power to turn back on. For obvious reasons, this individual (alone, or in aggregate), would never generate enough demand to fuel large-scale GI infrastructure, but these individuals are responsible for funding such efforts, and have unknowingly funded the GI that helps supply a solution to the loss of power problem. Another example would be a farmer concerned with water management. The farmer could fly an unmanned aerial system to collect information about water use on his or her property, but that GI would only help the individual farmer, failing to measure the impact of water use on other farms, downstream. However, an earth scientist with similar, aggregated data across multiple farms in a region could inform policy related to runoff or water rights. Specifically, this information might detail water use for individual farmers and their associated impacts regarding water availability on other farms and public lands. Much like any large-scale public service or good, an investment in SDIs must provide some value above and beyond the costs associated with having no GI at all.

National SDIs and the Global SDI (GSDI) present visions to find solutions to important national and international problems, but these large scale efforts carry less authority (and value) at the local levels, especially where governments and individuals are concerned. As a result, the value of GI cannot be market dependent. The efficient distribution equilibrium via price and rational acting parties would not lead to profits for many SDI purposes. More efficient use of natural resources, social infrastructure, and economic development, are all activities that increase the value of the GI when it is shared, but the ROI is difficult to measure. Any ROI study of SDIs would face challenges because some aspects of GI data collection require satellites, boats, planes, and equipment at high costs and without any immediately calculable value. Few entrepreneurs would be able to capitalize on the GI that results from much of geospatial data collected for government purposes (e.g., carbon measurements, or salinity, or even weather conditions). Still, many other GI can be extremely valuable for determining market areas, predicting real property values, and with inherent values that more than justify the costs and cost-recovery models for government GI distribution.

Other reasons for uniform and centralized data collection in SDIs include rigorous quality assurance and quality control—both of which add value to the GI. If sharing were not common through SDIs, then users and their communities would likely purchase more GI, spawning duplication and reducing reuse. Generally, these types of market conditions impose considerable search costs to determine sources of authoritative GI. This is especially true when multiple versions of the same geospatial data are made available from private data vendors, as opposed to government GI that presents itself as the authoritative, high quality source. To be blunt, GI quality is not like thumping a melon in the grocery store; GI quality requires more time to evaluate fitness for use even when properly described with formal metadata. To save search time, the status of the GI creators is often used by individuals to determine the value (Crompvoets, de Man, & Macharis, 2010). A trusted SDI reduces search costs for users and communities and removes the time to compare and contrast similar GI for fitness for use. Any ROI study would also need to measure these cost savings.

Authoritative GI creators like the USGS face a dynamic and changing market, especially as the Geoweb and its associated tools have emerged. As will be detailed in Chap. 6, the Geoweb introduces a different type of authority for GI. For example, much of the Geoweb data is crowd-sourced, especially from individuals with location-enabled mobile devices. This does not necessarily reduce the value of formalized GI, nor jeopardize its dominion as an authoritative source, but with the incorporation of more crowd-sourced, local knowledge and volunteered geographic information (VGI), SDIs can leverage more (and different) granularities of data than ever before. When users of GI are also GI creators, the vision of a global, real-time SDI seems more attainable than ever. Consider, for example, the use of VGI for the USGS's The National Map Corps (TNMCorps). It has volunteers successfully editing structure in all 50 states (http://nationalmap.gov/TheNationalMapCorps/). Similar to the way other crowdsourced information resources are populated, TNMCorps allows users to easily contribute GI. After basic quality control processes are conducted, this GI eventually becomes part of *The National Map*. Such efforts drastically increase the value of this NSDI, but the overhead (and price) of implementing this type of scheme is relatively low—promising fun, with volunteers receiving recognition badges. For now, the volunteers work on a relatively simple layer of municipal buildings but using citizen scientists and volunteers across SDIs present new stratum to ROI studies (Budhathoki & Nedovic-Budic, 2008).

Lastly, even if an SDI was eventually able to cover the planet to a level of detail equivalent to each blade of grass, any ROI study would not fully value that SDI. Advances in all fields are unpredictable and forecasting that accounts for the future use of data is highly speculative. For example, although many expected the predict the commercial implications, there was no way to predict the commercial implications that have been realized over the past 20 years. For example, when Amazon.com began operation in 1994, few would have predicted that by 2015 it would be the largest (and most valuable) retailer, as measured by market capitalization—surpassing Walmart (Kantor & Streitfeld, 2015). Again, the point here is that prophesizing the future value of any knowledge creation is problematic. Thus, more work is required to determine an equilibrium between SDI values and costs. These economic studies should successively inform the GI collected and shared. Fundamental work on the value of metadata would also inform the total SDI value, and some approaches are outlined in Chap. 5.

4.2.2 Users and Communities

With all the complexities associated with sharing GI, the SDI's purpose of meeting the demands of its users can get muddied. This is especially true when defining who or what uses the SDI is ultimately structured for. Consider some of the thematic content found in most national SDIs, such as *biodiversity and ecosystems*. A past president of the Society for the Preservation of Natural History Collections (SPNHC) provided a list of known biocollection users from health, public safety, agriculture, food security, conservation, ecotourism, genomics, recreational, government, commercial, NGOs, policy, and many others. When nearly anyone is a user, your community can appear to be ubiquitous. Still, different user groups and communities of practice have various information needs and unique human information seeking behaviors that should inform SDI design (topics which are covered in Chap. 8). Regardless of the particular theme, there is a similar breadth and depth of users and communities because government operations and the GI it provides impact so many facets of our political, economic, environmental, and cultural existence. Researchers and scientists working in any area need to share data, but with GI, a SDI may facilitate these processes. For example, a biologist could share their GI with a specimen portal like iDigBio as long as they complied with data ingestion requirements and guidelines (https://www.idigbio.org/). Sharing their GI in a systematic way allows access to that GI beyond any one institution or original creator. Certainly, no one person or organization could ever hope to collect global data on a particular species. Thus, sharing data has been a necessity to advance the study of biodiversity as it has in many other fields. However, suggesting that a lay person could successfully access SDIs and benefit from them is an untested assumption for many communities. In reality, SDIs are created *for* expert organizations *by* expert organizations (Craglia, 2007).

As alluded to previously, although stakeholders for open data could include all residents of a governed region, most residents are likely unaware of GI and how it impacts them. For example, in a 2005 UK study, only 41% of people living in a flood plain were aware of their situation. Worse, the majority of residents that were aware of their location on a flood plain did not believe a flood would impact them (Burningham, Fielding, & Thrush, 2008). In short, it should not be expected that all people will be literate with the details of GI or its potential implications. Thus, designing an SDI usable for all (individuals, organizations, and so forth) is unrealistic and unneeded.

Given the broad context presented of the SDI concept, its value, costs, users, and communities, the next section seeks to cover the necessary background of U.S. GI policy to set-up an in-depth NSDI discussion. The NSDI policy analysis summarizes milestones of the last thirty years stemming from the implementation of the NSDI and goes beyond discussion of the technological infrastructure that makes sharing GI possible.

4.3 Background of U.S. GI Policy

Peter Jefferson, an accomplished land surveyor, mapped the border between North Carolina and Virginia in 1749. Later, his son, Thomas Jefferson, proposed that states abandoned claims west of the Appalachian Mountains and drew the borders of what, after a series of Land Ordinances, became new U.S. states. As president, Thomas Jefferson purchased the Louisiana Territory, significantly increasing the size of the U.S. A key theme of expansion in the U.S. is the scientific demarcation of title to acquired lands. In his career, Jefferson attempted and failed to implement a national grid-based quadrant surveying scale but ultimately the Public Land Survey System (PLSS) was utilized to measure and map the growing country.

Survey work created lines and points to grid and divide lands with greater accuracy and precision than flawed European approaches of metes and bounds (e.g., *beginning at a stake and stones about forty feet from the center of the brook that runs across the road*). These efforts were made in order to assess the land for taxation in a consistent manner without the reliance on haphazardly measured metes or bounds that change (e.g., an oak tree at the property line dying).

In 1879, the USGS was created. It was formally charged with mapping the nation so that citizens could discover and use the vast resources of the U.S. Although the USGS was the first government agency to create and manage GI, it has been estimated that up to 80,000 agencies at various levels of government generate GI in the U.S., including the BLM (Bureau of Land Management), EPA (Environmental Protection Agency), NASA (National Aeronautics and Space Administration), NIMA (National Imagery and Mapping Agency), NOAA (National Oceanic and Atmospheric Administration), and the USDA (U.S. Department of Agriculture) (Masser, 1998; Tosta, 1997). Historically, the creation of the maps, not the data itself, drove policy for which agency did what. Until recently, the distinction between the outcome of GI creation (i.e., print map) and the geospatial data that enables that outcome proves problematic and indistinguishable in GI policy.

As geospatial data proliferated over the past several decades, and as the number of GIS users grew, the necessity of sharing the generated GI to address global problems became clearer. In turn, a vision of a digital earth emerged in U.S. policy in 1990 (US OMB). Again, the core idea for such movements is to help citizens and local development authorities prosper from the information. These social, economic, and environmental implications drive modern GI policy in all nations, but the digital earth concept expands beyond the idea of local and regional land management to one that includes a more global perspective requiring the construction of SDIs.

The milestones in Table 4.1 relate to the construction of the NSDI and help structure the following sections. This oversimplification of milestones fails to acknowledge the efforts of individual agencies at all levels of government and additional entities outside the federal public sector, but provides a succinct overview for policy analysis. Ultimately, the NSDI vision is a continuing manifestation without end and the infrastructure is the accumulated actions of any professional creating or curating government GI (Rhind, 1999).

4.3.1 Building a Digital Earth

Prior to the regulations outlined in the OMB's Circular A-16 (1990), *Coordination of Surveying, Mapping, and Related Spatial Data Activities*, few regulations existed to govern the organization and dissemination of digital geospatial data collected by the multiple agencies of the U.S. In response to a nationwide move from print to digital GI for agencies and users, Circular A-16 called for the creation of an interagency committee, the Federal Geographic Data Committee (FGDC) to address the nation's digital geospatial data needs. The FGDC was intended to "[promote] the

Table 4.1. Timeline of US NSDI milestones

Year	US NSDI milestones
1990	FGDC was created to coordinate the development, use, sharing, and dissemination of geospatial data in Circular A-16
1993	US MSC report *Towards a coordinated spatial data infrastructure for the nation* was released to call for the development of a NSDI
1994	EO 12906, *Coordinating geographic data acquisition and access: the National Spatial Data Infrastructure* charged the FGDC with the creation of the NSDI
2002	Revised Circular A-16, called for geospatial data from all levels of government, the private sector, and academia to be widely integrated and the creation of a web-based portal to access US geospatial data
2003	Geospatial One-Stop launched
2004	FGDC released the *NSDI Future Directions Initiative: Towards a National Geospatial Strategy and Implementation Plan* to refocus the NSDI policy on collaboration
2006	Geo.data.gov launched
2010	Revised Circular A-16, called for adjusting guidance for the FGDC in identifying geospatial themes
2014	Geoplatform.gov launched

coordinated development, use, sharing, and dissemination of geospatial data on a national basis" (US OMB 1990). In order to create support among governmental stakeholders and to raise the political visibility of the FGDC, the cadre of committee members were selected from each organization's policy level leaders (Masser, 2005a).

Unrelated to the FGDC, but also resulting from the changes in GI organization, access, and use in government operations, the National Research Council's U.S. Mapping Science Committee (MSC) released a report entitled *Toward a Coordinated Infrastructure for the Nation* in 1993 (National Research Council, U.S.. Mapping Science Committee, 1993). In the report, the MSC explored the impact of GI that could be aggregated, transformed, stored, reused, and shared across agencies. The MSC also introduced a "top down" model of base map creation for agencies (NRC, 1993, p. 8). The report suggested the creation of an infrastructure for the nation's geospatial data, similar to those that managed other national infrastructure assets (e.g., the Federal Highway Administration) at the time. In particular, it was recommended that all geospatial data for the nation be assembled, including a collection of the "materials, technology, and people necessary to acquire, process, and distribute" that GI (NRC, 1993, p. 16). The MSC also recongnized the existence of an unregulated and incoherent *ad hoc* SDI, but called for the creation of a coordinated NSDI designed to facilitate the use of geospatial data among stakeholders.

To this end, President William Jefferson Clinton issued Executive Order No. 12906 (EO 12906, Clinton, 1994), *Coordinating Geographic Data Acquisition and Access: The National Spatial Data Infrastructure*. This order stressed the importance of coordinated development, use, sharing, and dissemination of geospatial data on a national basis and called for the formal creation of a coordinated NSDI (Clinton, 1994). EO 12906 defined the NSDI as "the technology, policies, standards, and human resources necessary to acquire, process, store, distribute, and improve utilization of geospatial data" (Clinton, 1994). NSDI policy goals included reducing the duplication of work among agencies and increasing the interoperability between agencies. These two goals allowed for a more coordinated dissemination and storage of government GI for the NSDI to be realized. The vision of the NSDI was defined in EO 12906 "to promote economic development, improve our stewardship of natural resources, and protect the environment" (Clinton, 1994). Coordinated GI collection and sharing affect the quality of a variety of public and private sector decision-making abilities, and the policy reverbs the benefits chord of the triple bottom line: economic, societal, and environmental. The embedded potential to reduce government work duplication appealed to many policy makers and the NSDI only faced implementation challenges like interoperability. The dream of sharing all GI across the government was simple, but given the multitude of datums, projections, data formats and types, technological changes, and other limitations, the implementation of this vision has proven difficult.

Although challenges remain, the existence of a more coordinated NSDI assists in decision-making for natural resources and economic development. In addition, the more efficient and effective coordination of GI saves agencies both time and money by partially alleviating the burden of individual geospatial data collection and maintenance (Federal Geographic Data Committee, 1997). The costs of acquiring and processing base map elements is extremely expensive. Thus, it was believed that by removing the duplication of work and implementing a more defined National Spatial Data Framework, the U.S. could significantly reduce agency costs (Federal Geographic Data Committee, 2004). Specifically, greater collaboration among stakeholders results in more efficiency and increased savings. In addition, sharing GI improves data quality by increasing the number of individuals with the ability to find and correct errors beyond traditional quality control and quality assurance processes. The reduction in repetitive GI collection further allows individuals more time to work on other tasks (Federal Geographic Data Committee, 1997).

Ultimately, the newly formed FGDC was charged with the initial tasks related to the NSDI's development (Clinton, 1994). Two of these FGDC tasks were the establishment of a National Geospatial Data Clearinghouse and the creation of a National Digital Geospatial Data Framework (Clinton, 1994). These two broad tasks can be further broken down into three separate areas of focus for the FGDC—(1) the National Geospatial Data Clearinghouse (i.e., the means of disseminating for GI access and use), (2) a digital spatial data framework (i.e., the coordinated infrastructure for the nation's agencies to collaborate), and (3) metadata standards to enable these two tasks to be achieved most effectively (i.e., information organization) (Masser, 2005b). The metadata standards receive fuller exploration in Chap. 5, but discussion of the first two FGDC tasks appear in the following sections. The NSDI framework and clearinghouse mirror each other from inception to this day. The types of federal data available in clearinghouses result from the plan of the frameworks presented, but over time, new policies and additional advances in GI have altered both of them.

4.3.2 A Clear Framework

The National Spatial Data Framework is an ongoing collaborative effort created by the FGDC that regulates the development, maintenance, and integration of public and private geospatial data over geographic areas by county, region, state, and federal agencies (Federal Geographic Data Committee, 1997). The NSDI failed to demarcate a clear federal boundary and the original framework seemed to have no jurisdictional bounds. This proved problematic for several reasons. First, elements of GI have a global scale and some government GI, while other GI comes from local and state agencies. Further, much of data collected by local and state agencies is/was not collected with aggregation in mind. As a result, repurposing these data for use at the federal level is not necessarily straightforward, nor does it lend itself for transformation into a uniform data source for an entire country. That said, the original framework still adhered to a "top down" view for organization, and used seven common digital geospatial data themes: (1) geodetic control, (2) orthoimagery, (3) elevation, (4) transportation, (5) hydrography, (6) governmental units, and (7) cadastral information. The general framework also provided guidelines for sharing and using GI, and encouraged best practices for the maintenance of geospatial data (Federal Geographic Data Committee, 1997).

To coordinate the collection and maintenance of each of the seven most common digital spatial data themes, federal government agencies were assigned responsibility for the themes that most closely matched their organizational purposes. For example, the National Geodetic Survey took primary responsibility for geodetic control, which creates common reference systems and allows for established coordinate positions of all other geospatial data. The USGS handled orthoimagery (e.g., remotely sensed and georeferenced images of earth), elevation (e.g., the georeferenced vertical position of objects above or below a datum's surface), and hydrography (e.g., surface water features). All transportation networks (e.g. railroads) were the responsibility of the Department of Transportation. The U.S. Census Bureau managed the political boundaries of governmental units. Cadastral spatial data (e.g., history and current information about land parcels) were handled by the Bureau of Land Management. Without a well-organized framework, access and sharing through a clearinghouse would be chaotic. Unfortunately, the agency-specific approach proved too simplistic to work, in practice, as GI generated by the U.S. agencies do not precisely reflect the structure of government or its GI needs.

In 2010, the OMB revised Circular A-16 for a second time, *Coordination of Geographic Information and Related Spatial Data Activities* to present principles to adjust guidance for the FGDC in identifying geospatial themes (US OMB, 2010). There were two important outcomes from this process. First, the revision required GI themes to be national in scope. Second, because government GI does not neatly match up with only *one* federal agency, a new assessment of GI needs and assets for each agency was mandated. Specifically, the revised Circular A-16 required a portfolio of all GI in circulation. The result was the creation of sixteen National Geospatial Data Asset (NGDA) Theme Communities that cut across multiple agencies. These communities included: Biodiversity and Ecosystems, Cadastre, Climate and Weather, Cultural Resources, Elevation, Geodetic Control, Geology, Governmental Units, and Administrative, and Statistical Boundaries, Imagery, Land Use-Land Cover, Real Property, Soils, Transportation, Utilities, Water—Inland, Water—Oceans & Coasts (Federal Geographic Data Committee, 2014). Embedded within these themes, there are many pragmatic reasons as to why different groups collect similar (or identical) geospatial data, but for the NSDI to be successful, it requires a deeper understanding of why there is duplication and how much these duplicative efforts cost. The National Geospatial Data Asset Management Plan reiterated the reasons for creating a federal portfolio; (1) increase shared use; (2) increase the value of data as more will rely on it; (3) increase opportunities for partnering in the data lifecycle; (4) commodify the return on investment of all the GI; and (5) identify gaps (Federal Geographic Data Committee, 2014).

If readers still find the overarching GI framework confusing, here is a borrowed analogy from the USGS that may help readers think spatially about the policy implementation of the NSDI. The digital earth that would result from a compilation of all data themes in the NSDI framework would have many layers from the USGS *The National Map* (National Research Council, U.S., 2003). *The National Map* includes orthoimagery (aerial photographs), elevation, geographic names, hydrography, boundaries, transportation, structures, and land cover and is a significant piece of the NSDI (http://nationalmap.gov/). Much like the larger NSDI, this huge piece of the SDI puzzle, has the triple bottom line objective to "assist in public safety during natural and human-induced disasters, assure the effective use and protection of the nation's resources, and support literally hundreds of applications that contribute to every citizen's daily life" (National Research Council, U.S., 2003, p. 20). Federal GI comes in two formats—blankets and patchwork quilts. A blanket coverage would be analogous to all the themes which exist at national levels (or global scales) that are created with the same spatial resolution like a solid sheet spanning the entire nation or world collected at the same scale (National Research Council, U.S., 2003). Some blanket examples are the Digital Orthophoto Quadrangle (DOQ) or Bureau of the Census's Topologically Integrated Geographic Encoding and Referencing (TIGER) files.

While the blanket metaphor only works for some themes (e.g. Geodetic Control), the patchwork quilt metaphor fits better with the other themes that require a greater variety of contributors. Patchwork quilt examples include regional (e.g., Lower Cumberland Light Detection and Ranging (LiDAR)), state (e.g., Michigan Drinking Water Wells), or local (e.g., Manatee County Garbage Routes) data. For a myriad of reasons, these GI are not collected across the nation. Many "bottom up" GI themes such as Biodiversity or Cadastre could be weaved together from a variety of smaller administrative (and areal) units into one national layer. Technology should not be a barrier to this process. Given time, government funding, and user demand, the collection of local GI that requires ground truthing and local knowledge for accurate creation *could* be organized, accessed, and used. For instance, the 3D Elevation Program (3DEP) is one example of how growing demand for high-quality topographic data drives the completion of another nationally complete patchwork quilt (http://nationalmap.gov/3DEP/). Basically by 2024, the program should result in a LiDAR coverage for the conterminous U.S., Hawaii, and the U.S. territories. The importance of LiDAR and its use for spatial analysis means that a federal option was the preferred pathway to completion—ensuring quality control and widely considered a worthy investment. Work like the 3DEP project clearly helps stitch together sections of *The National Map*, and subsequently the NSDI. However, for data concerning garbage routes, market demand remains low and that type of information remains sporadically available at the local level. Conversely, the demand for cadastral information is high, but without a maintained federal government dataset, many cadastral databases are difficult to acquire. Some counties charge exorbitant fees for distributing these data, while others make it freely available. In short, because not all GI are valued equally, or invested in evenly, patchwork quilts easily outnumber the "blankets" in the U.S.

4.3.3 A Framed Clearinghouse

The first revision to Circular A-16 occurred in 2002. In addition to updating its name (*Coordination of Geographic Information and Related Spatial Data Activities*), the revision called for more collaboration between federal, state, and local governments, private sector stakeholders, and academicians (US OMB, 2002). This renewed emphasis on collaboration was part of a wider federal government movement toward openness, enabled by web-based services (E-Government Act of 2002, 2002). For example, the E-Government Act of 2002 emphasized the government's focus on utilizing information technologies and specifically called for "Internet service delivery for more accessible, responsive, and citizen-centered government" (Tang & Selwood, 2005, p. 34). Of note is that the revision to A-16 also introduced the exclusion of classified national security-related spatial data activities from the Departments of Energy and Defense, the intelligence agencies, as well as GI from tribal governments not paid for by federal funds. It is important to reiterate that the systematic exclusion of data is an interesting choice for the U.S. government. As detailed earlier, when looking to balance privacy, national security, and openness, there is a long history of interesting and/or controversial decisions made for data in the U.S.

One of the more important e-government initiatives realized in 2003 was Geospatial One-Stop (GOS) (www.geodata.gov), a web-based portal to make government GI available. The creation of GOS presented the first attempt at a truly National Geospatial Data Clearinghouse. The FGDC hired the Environmental Systems Research Institute (Esri) to develop the geoportal that became GOS (Tang & Selwood, 2005). Again, the basic idea of a clearinghouse for GI is the seamless integration of searches across regional, administrative, and organizational boundaries. That said, many information professionals may wonder why a clearinghouse is required when so much information can be found simply through search engines. There are several reasons for this. First, if every GI creator served up geospatial data online, with metadata, then the data would be discoverable on a case by case basis. However, by putting the nation's geospatial assets in one clearinghouse, the collocation of GI across themes aids serendipity in search and discoverability so users can locate similar resources that they may (or may not) have discovered otherwise. Third, the clearinghouse also empowers users to locate all the data related to a particular place or theme. Unlike text or images, or even web-based mapping applications, GI has a multitude of unique metadata components and complex data formats (as detailed in Chaps. 2 and 3). This variety and volume are not easily handled by the functionality of search engines, especially since they are purpose-built for finding other types of information (e.g., text, images, and so forth). Advanced queries that take into account the special nuances of GI, including spatial coordinates, scales, and data formats, are required to intelligently and efficiently retrieve GI (Kerski & Clark, 2012). Therefore, a central National Geospatial Data Clearinghouse remains necessary to fulfill the GI access and use tasks of NSDI policy.

It is important to note that there was additional functionality built into the GOS. For example, in addition to providing standard, text-based/string queries, it allowed users to conduct spatial queries on a map (i.e., search by geography (where) without the use of text). The GOS service also permitted users to search based on topical classification (what) and time (when). In addition to locating GI, GOS provided standardized access to GI from static files through file transfer protocol (ftp) or via other web services. Unfortunately, as producer and user needs evolve, all technologies of this ilk have an indeterminate shelf-life. In the case of the GOS, it was retired in 2011. However, it is important to remember that at the time of its debut, GOS included many of the functions taken for granted in today's Geoweb tools—the ability to view maps, download GI, from the state, local, and tribal governments, as well as the private sector and academia. In recognition of its usability, use, and ability to coordinate across agencies, GOS received an Excellence.Gov award in 2006 (The White House, Office of the Press Secretary, 2006). The data was later made accessible via geodata.gov in 2006 until 2010 when federal GI could be discovered through Data.gov. In 2014, the official NSDI one-stop clearinghouse was launched as the GeoPlatform Catalog. The GeoPlatform Catalog works seamlessly with Data.gov and provides search, documentation, and API for various theme developer communities. Having one clearinghouse that supports open formats, data standards, common core and extensible metadata, as well as the promise to allow users to share code moves this part of the FGDC task closer to a singular federal access point for the nation's GI rather than a long list of data catalogs. Some key GI products used to help in emergency operations have also been listed in the GeoPlatform through Homeland Security's Geospatial Concept of Operations (GeoCONOPS).

4.3.4 Open Government

Complying with previous legislation, much of the original NSDI vision assumed that sharing GI across government agencies and beyond government operations was a good thing. At the same time, the NSDI policy expressly ensured the protection of the privacy and security of citizens' personal data, which aligned with other facets of U.S. legislation. To be more specific, the NSDI policy protected the proprietary interests of stakeholders, including central and local governments, the commercial sector, not-for-profits, academics, and individuals, in the creation of the NSDI. The NSDI continues to serve as a meta-information policy for the U.S.—all of it based on an executive order and one OMB Circular. That said, many other policies will have an impact on its continued implementation (e.g., E-Government Act of 2002, 2002). Most recently the EO 13642, *Making Open and Machine Readable the New Default for Government Information*, stipulated that newly-generated government data are now required to be made available in open, machine-readable formats while continuing to ensure privacy and retain security (Executive Order No. 13642, 2013). This includes government GI and reemphasizes the spirit of prior government information policy.

With the most recent revisions to Circular A-16, new framework themes, and the latest clearinghouse, the impact of openness should cement greater access to GI for citizens and organizations into the foreseeable future. As policies change and clarify the NSDI scope, its success relies on metadata creation and standards compliance. In this respect, fulfillment of the NSDI's vision relies both on the hopes that future policy revisions reflect increased openness and at the mercy of the minuscule building blocks of this cyberinfrastructure (e.g., metadata). During the writing of this book, the most recent legislative action was the introduction of Geospatial Data Act of 2015 on March 16, 2015. The bill was referred to the Committee on Commerce, Science, and Transportation. It reaffirms vows of all prior policy actions, with clearer wording about "free access for the public to geospatial data, information, and interpretive products" and new "support and advance the establishment of a Global Spatial Data Infrastructure" (https://www.congress.gov/bill/114th-congress/senate-bill/740/text). As of June 1, 2016, the bill had nine co-sponsors (https://www.govtrack.us/congress/bills/114/s740). Realistically, information policy that is as technically complex and as crosscutting as the NSDI will continue to evolve through other federal government actions, helping pursue nation's triple bottom line for GI—"to promote economic development, improve our stewardship of natural resources, and protect the environment" (Clinton, 1994). The remaining portion of this chapter focus on considerations for all SDI related to pricing, security, and privacy.

4.4 SDI Considerations for Pricing, Privacy, and Security Moving Forward

All current government SDIs are classic examples of "top down", centralized control by government. In this respect, the NSDI and others like it resemble article 71, item 'p' of the pre-1989 Soviet Union's Russian Constitution that reserved the responsibility of geodesy, cartography, and the naming of geographical features to the federal government (Rhind, 1999). As outlined previously, in some instances, this centralized, hierarchal structure may be cost effective and necessary for national or global implementation. In addition, the power state and/or intergovernmental organization facilitates the information needs that drive GI standardization in organization, access, and use. For these reasons and as a result of other historical precedents related to federal GI collection (e.g., the U.S. Constitution's requirement of the Census), the basis exists in the U.S. and elsewhere for centralized NSDI operations (Masser, 1998). This "top down" precedent for SDIs exists in all national level information policies, but it is worth acknowledging that other structuring is possible. In an environment where big data, VGI, and other trends may motivate a call for a reorganization of authority on some types of GI, governments need to consider what GI remains within their purview and what GI can they afford to organize, make accessible, and usable. Unlike centralized control, other aspects of GI policy both in the U.S. and elsewhere require further examination—pricing, privacy, and security.

The NSDI's leadership role led to a National Spatial Data Framework, the current National Spatial Data Clearinghouse, and FGDC metadata standards. To continue the NSDI or develop any government SDI requires not only cooperation among various levels of government and across stakeholders but upkeep and updating of any existing infrastructures. In the spirit of democracy, GI organization, access, and use requires continuous investment in those infrastructures. Many users outside of governments (e.g., academicians, private citizens, and private sector stakeholders) will need access to government GI for various GI uses. With increased demand for GI, private sector geospatial technologies exist to meet users' demands and many commercial geodata portals exist to cover value added costs. While the private sector facilitates production, cleaning, and dissemination of GI in many ways, government GI at all levels of government would benefit the most citizens if it were as freely available as possible to ensure equity and accountability in democracies for non-classified GI (Rhind, 1999).

If we assume the government's role is to minimize the overall costs to citizens and maximize the equality of access, there are questions as to how the government would price GI. The balance between how services can be most equitably distributed and how cost can be most efficiently managed are at the heart of all pricing considerations. The FGDC mandated that agencies cannot charge amounts higher than those necessary to recover the cost of dissemination by those agencies to geospatial data users (FGDC, 1997). This precedent follows prior policy from the Independent Office Appropriations Act of 1952, which suggests that cost recovery models are appropriate and requires that agencies not charge beyond the costs of dissemination (Independent Offices Appropriations Act of 1952, 1952). There are several arguments for zero cost dissemination of GI. First, some feel any new charge is a second charge (e.g., taxpayers fund the GI creation, but not the dissemination). Second, the total gains of cost recovery revenues may not be fiscally responsible if the additional costs of collecting fees incur new costs (e.g., library book fine enforcement actually costs more than it typically collects). Third, the outlined intangible benefits and unmeasured values that widespread use of freely available GI is probably a good thing (Rhind, 1999). Further reductions in the cost of data transfer have been realized as agencies moved from dissemination in print to CDROM to online downloads (Taylor, 1998). With twenty additional years of technological advancements, dissemination costs are the same as sharing any GI on existing Geoweb infrastructures. During the writing of this book, a substantial amount of U.S. government GI is served up through an Esri powered ArcGIS Online (AGOL) interface at the GeoPlatform.gov Map Viewer. Another GeoPlatform tool is called the Marketplace. It is structured to make FGDC member agencies aware of other GI available from other FGDC members. Again, the GeoPlatform.gov Catalog serves as the NSDI clearinghouse and users may find the types of GI available using it. Regardless of the platform, GI access and use depends on pricing—and in this instance, data are freely accessible to all FGDC members.

Despite these near zero dissemination costs for GI, cost recovery is practiced by many at other levels of government beyond the FGDC members. In all cases, cost recovery is self-policed. Although we are optimistic that governments will see the advantages of sharing GI and will take steps to facilitate this process through geospatial technologies, there remain many local and state government agencies that charge for GI access and use. In many cases citizens that would like to view municipality GI may use geobrowsers, but if citizens want to download the data it is not always free. Sometimes the GI needs to be purchased when a municipality is using a cost recovery model and in some instances the GI is not available at all. To some extent, these restrictive pricing practices to GI access and use help address security concerns, albeit awkwardly. Still, no systematic study of government GI pricing exists. This topic is worthy of future research. To be clear, the pricing considerations for operating an SDI will remain problematic as long as each agency is allowed to determine cost recovery pricing schemes. Historically GI creation and curation costs remained an expensive activity that only centralized governments could afford to control. However, this is changing with commercial geoservices. A short-term fix might include extending FGDC policy to other SDIs, helping to clarify cost recovery models for GI access and use, but again, more research is needed in this domain.

Once again, it must be noted that privacy and security restrictions exist for good reasons. There is no circumstance or situation where these restrictions and safeguards should be sold. These personal and defensive considerations restrict GI access and use and are much clearer in newer NSDI policy. That said, it is important to note that both current and legacy policy instruments, globally, impact GI organization, access, and use. For example, in the U.S., a number of policies have been highly influential, including The Paperwork Reduction Act, The Government Paperwork Elimination Act of 1999, The Government Performance and Results Act of 1993, The Federal Records Act, OMB Circular A-130, OMB Circular A-119, The Freedom of Information Act and the Electronic Freedom of Information Act Amendments to 1996, The Privacy Act, The Clinger-Cohen Act of 1996, The Stafford Act, Federal Acquisition Regulations, among others. Moving forward, SDIs will need to collect data that is not shared. Future policy actions may address concerns over what GI is collected by whom and for what purposes. Given that "Big Data sensors, tools, and applications are in the hands of powerful institutions rather than ordinary people" (Crampton, 2015, p. 521), it must be recognized that privacy and security are not inverse concepts. The same tools that enable amazing location-based services that make everyday life easier also allow for mobile devices and their users to be tracked and their behavior observed. As a result, future GI policy considerations for any SDI need to codify the shifting boundaries for both privacy and security.

4.5 Conclusion

This chapter provided an overview of GI policy with a focused review of the SDI concept. In the U.S., the NSDI policy attempts to realize the dream of a digital earth and the analysis in this chapter points to successes and future considerations for further implementation. Despite advancements in various Geoweb tools and virtual globes, increased compliance with FGDC metadata standards across government GI, greater participation in national clearinghouses, a clearer framework to determine all U.S. GI assets, and additional contributions by non-FGDC member government entities, private sector and academic organizations, the overall success of GI organization, access, and use relies on the individual GI creators and GI curators. Ironically, the lack of knowledgeable workers with Information Science training leads to more costly GI creation. The NSDI seeks to reduce duplication of work and increase sharing, and to accomplish those tasks it is important that policy leaders in the U.S. and elsewhere support GI organization, access, and use education. In addition, more research needs to be done to address long-term digital preservation issues of GI within these SDIs (Goodchild et al., 2012).

References

Bishop, W., Haggerty, K., & Richardson, B. (2015). Usability of E-government mapping applications: Lessons learned from US National Atlas. *International Journal of Cartography, 1*(2), 1–17.

Budhathoki, N. R., & Nedovic-Budic, Z. (2008). Reconceptualizing the role of the user of spatial data infrastructure. *GeoJournal, 72*(3–4), 149–160.

Burningham, K., Fielding, J., & Thrush, D. (2008). 'It'll never happen to me': Understanding public awareness of local flood risk. *Disasters, 32*, 216–238. doi:10.1111/j.1467-7717.2007.01036.x.

Carver, S. (2003). The future of participatory approaches using geographic information: Developing a research agenda for the 21st century. *URISA Journal, 15*(1), 61–71.

Chester, J. (2007). *Digital destiny: New media and the future of democracy*. New York: New Press.

Clinton, W. (1994, April 13). *Coordinating geographic data acquisition and access: The National Spatial Data Infrastructure*. Executive Order 12906. Retrieved from http://govinfo.library.unt.edu/npr/library/direct/orders/20fa.html.

Craglia, M. (2007). *Volunteered geographic information and spatial data infrastructures: When do parallel lines converge?* Retrieved from http://www.ncgia.ucsb.edu/projects/vgi/docs/position/Craglia_paper.pdf.

Crampton, J. W. (2015). Collect it all: National security, big data and governance. *GeoJournal, 80*(4), 519–531. doi:10.1007/s10708-014-9598-y.

Crompvoets, J., de Man, E., & Macharis, C. (2010). Value of spatial data: Networked performance beyond economic rhetoric. *International Journal of Spatial Data Infrastructures Research, 5*, 96–119.

Dunn, C. E. (2007). Participatory GIS—A people's GIS? *Progress in Human Geography, 31*(5), 616–637.

E-Government Act of 2002. (2002). H.R. 2458/S. 803, 107th Cong. § 2.

Elkington, J. (1998). *Cannibals with forks: The triple bottom line of 21st century business*. Gabriola Island, BC: New Society Publishers.

Ellison, N., Steinfeld, C., & Lampe, C. (2007). The benefits of Facebook "Friends": Social capital and college students' use of online social network sites. *Journal of Computer-Mediated Communication, 12*(4), 1143–1168. doi:10.1111/j.1083-6101.2007.00367.x.

Erik de Man, W. H. (2006). Understanding SDI; complexity and institutionalization. *International Journal of Geographical Information Science, 20*(3), 329–343. doi:10.1080/13658810500399688.

Evans, B. J. (2011). Much ado about data ownership. *Harvard Journal Law and Technology, 25*(1), 69–130.

Exec. Order No. 13642. (2013). 3 C.F.R.

Federal Communications Commission [FCC]. (2010). *Connecting America: The national broadband plan*. Retrieved from: http://www.fcc.gov/national-broadband-plan.

Federal Geographic Data Committee. (1997). *Framework introduction and guide*. Retrieved from http://www.fgdc.gov/framework/handbook/index_html.

Federal Geographic Data Committee. (2004). *NSDI future directions initiative: Towards a national geospatial strategy and implementation plan*. Retrieved from http://www.fgdc.gov/policyandplanning/future-directions/reports/FD_Final_Report.pdf.

Federal Geographic Data Committee. (2014). *National geospatial data asset management plan*. Retrieved from https://www.fgdc.gov/policyandplanning/a-16/ngda-management-plan.

Goodchild, M. F., Guo, H., Annoni, A., Bian, L., Bie, K. D., Campbell, F., et al. (2012). Next-generation digital earth. *Proceedings of the National Academy of Sciences, 109*(28), 11088–11094. doi:10.1073/pnas.1202383109.

Grubesic, T. H. (2012a). The US national broadband map: Data limitations and implications. *Telecommunications Policy, 36*(2), 113–126.

Grubesic, T. H. (2012b). The wireless abyss: Deconstructing the US national broadband map. *Government Information Quarterly, 29*(4), 532–542.

Grubesic, T. H. (2016). Future shock: Telecommunications technology and infrastructure in regional research. In R. W. Jackson & P. V. Schaeffer (Eds.), *Regional research frontiers: The next 50 years*. New York: Springer.

Grubesic, T. H., & Mack, E. A. (2015). *Broadband telecommunications and regional development*. New York: Routledge.

Hendriks, P. H., Dessers, E., & Van Hootegem, G. (2012). Reconsidering the definition of a spatial data infrastructure. *International Journal of Geographical Information Science, 26*(8), 1479–1494. doi:10.1080/13658816.2011.639301.

Higgins, S. (2008). The DCC curation lifecycle model. *International Journal of Digital Curation, 3*(1), 134–140 Retrieved from http://www.ijdc.net/index.php/ijdc/article/viewFile/69/48.

Independent Offices Appropriations Act of 1952. (1952). 82nd Cong. § 2.

Jaeger, P. T., Shneiderman, B., Fleishmann, K. R., Preece, J., Qu, Y., & Wu, P. F. (2007). Community response grids: E-government, social networks, and effective emergency management. *Telecommunications Policy, 31*, 592–604. doi:10.1016/j.telpol.2007.07.008.

Kantor, J., & Streitfeld, D. (2015). Inside Amazon: Wrestling big ideas in a bruising workplace. *New York Times*. Retrieved from http://tinyurl.com/o2vdvtf.

Kapucu, N. (2004). Interagency communication networks during emergencies: Boundary spanners in multiagency coordination. *American Review of Public Administration, 36*, 207–225. doi:10.1177/0275074005280605.

Kerski, J., & Clark, J. (2012). *The GIS guide to public domain data*. Redland, CA: Esri Press.

Masser, I. (1998). *Government and geographic information*. London: Taylor & Francis.

Masser, I. (2005a). *The future of spatial data infrastructures*. Paper presented at ISPRS workshop on service and applications of spatial data infrastructure, XXXVI (4/W6), Hangzhou, China. Retrieved from http://www.commission4.isprs.org/workshop_hangzhou/papers/7-16%20Ian%20Masser-A001.pdf.

Masser, I. (2005b). *GIS worlds: Creating spatial data infrastructures*. Redlands, CA: Esri Press.

McClure, C. R., & Jaeger, P. T. (2008). Government information policy research: Importance, approaches, and realities. *Library and Information Science Research, 30*(4), 257–264. doi:10.1016/j.lisr.2008.05.004.

Mossberger, K., Tolbert, C. J., & McNeal, R. S. (2008). *Digital citizenship: The internet, society, and participation*. Cambridge, MA: MIT Press.

Nation Communications and Information Administration. (2000). *National Telecommunications and Information Administration Annual Report*. Retrieved from https://www.ntia.doc.gov/report/2001/ntia-2000-annual-report-congress.

National Research Council (U.S.). Mapping Science Committee. (1993). *Toward a coordinated spatial data infrastructure for the nation*. Retrieved from http://www.nap.edu/read/2105/chapter/1.

National Research Council (U.S.) (2003). *Weaving a national map: Review of the U.S. Geological Survey concept of the national map*. Washington, DC: The National Academic Press.

Obermeyer, N. J., & Pinto, J. K. (2008). *Managing geographic information systems* (2nd ed.). New York: Guilford Press.

Organisation for Economic Co-operation and Development (OECD). (n.d). OECD broadband portal [Web portal]. Paris, France: OECD. Retrieved February 17, 2010, from http://www.oecd.org/sti/ict/broadband.

Peng, Z. R. (2005). A proposed framework for feature-based geospatial data sharing: A case study for transportation network data. *International Journal of Geographic Information Science, 19*(4), 459–481.

Prieger, J. E., & Hu, W.-M. (2008). The broadband digital divide and the nexus of race, competition, and quality. *Information Economics and Policy, 20*(2), 150–167. doi:10.1016/j.infoecopol.2008.01.001.

Priest, D., & Arkin, W. M. (2010). *Top Secret America—A Washington Post investigation. A hidden world, growing beyond control: The government has built a national security and intelligence system so big, so complex and so hard to manage, no one really knows if it's fulfilling its most important purpose: Keeping citizens safe*. Washington Post, pp. 1. Retrieved from http://secure.afa.org/edOp/2010/Washington_Post_Intelligence_Series.pdf.

Privacy Act of 1974. (1974). P.L. 93–579

Rajabifard, A., & Williamson, I. P. (2001). *Spatial data infrastructures: Concept, SDI hierarchy and future directions*. Proceedings of GEOMATICS'80 conference, Tehran, Iran.

Rhind, D. (1999). National and international geospatial data policies. In P. A. Longley, M. F. Goodchild, D. J. Maguire, & D. W. Rhind (Eds.), *Geographical information systems—Principles and technical issues* (Vol. 1, pp. 767–787). New York: Wiley.

Sunstein, C. R. (1986). Government control of information. *California Law Review, 74*(3), 889–921.

Tang, W., & Selwood, J. (2005). *Spatial portals: Gateways to geographic information*. Redlands, CA: Esri Press.

Taylor, D. R. (1998). *Policy issues in modern cartography*. Oxford, England: Elsevier Science Ltd..

Theraulaz, G., Gautrais, J., Camazine, S., & Deneubourg, J. L. (2003). The formation of spatial patterns in social insects: From simple behaviours to complex structures. *Philosophical Transactions of the Royal Society of London A: Mathematical, Physical and Engineering Sciences, 361*(1807), 1263–1282.

Tosta, N. (1997). Data revelations in Qatar: Why the same standards won't work in the United States. *GeoInfo Systems, 7*(5), 45–49.

USA PATRIOT Act of 2001. (2001). P.L. 107–56.

U. S. Office of Management and Budget. (1990). *Coordination of surveying, mapping, and related spatial data activities*. Retrieved from https://www.whitehouse.gov/omb/circulars_a016/.

U. S. Office of Management and Budget. (2002). *Coordination of geographic information and related spatial data activities*. Retrieved from https://www.fgdc.gov/policyandplanning/a-16/circular-A-16.pdf.

U. S. Office of Management and Budget. (2010). *Coordination of geographic information and related spatial data activities*. Retrieved from https://www.whitehouse.gov/sites/default/files/omb/memoranda/2011/m11-03.pdf.

The White House, Office of the Press Secretary. (2006). Department of Interior and USGS Receive Excellence. Gov Award for Geospatial One-Stop. Award-Winning Portal Provides GIS Services, Applications, and Rich Data Content [Press release]. Retrieved from http://georgewbush-whitehouse.archives.gov/omb/egov/press_releases/gtog/060412_geospatial.html.

Williamson, I., Rajabifard, A., & Binns, A. (2006). Challenges and issues for SDI development. *International Journal of Spatial Data Infrastructures Research, 1*, 24–35.

Chapter 5
Metadata

Abstract This chapter presents an introduction to information representation of geospatial data and cartographic resources. Metadata serves as the scaffolding that allows users to find, identify, select, and obtain geographic information (GI). These organizational concepts facilitate both GI access and use by representing information objects with specific approaches. Metadata as a topic unto itself, without domain specific syntax and technical jargon, gives readers a broad view of issues that cross disciplines, sectors, and communities. This chapter includes a section of the terminology and tools of knowledge organization, as well as covering other key concepts. A review of the ever-changing schemas, profiles, and standards of GI metadata provides current best practices.

5.1 Metadata Are

Metadata are data about data, but this definition fails to disambiguate the term from any other neologism because the term finds new meanings as different people use it across domains. Metadata encompass anything about the data, including the structure of the data, the producers of the data, the spatial or temporal nature of the data, and any number of other elements related to information representation. Without the use of more exacting terms, which coincidentally do exist within the domain of Information Science, metadata discussions quickly descend into semantic and syntactic bootstrapping. For example, consider the South Park episode *Starvin' Marvin in Space,* in which Marklars on the planet Marklar use the word *marklar* to refer to any place, person, idea, concept, or any other noun (Brady, Parker, & Stone, 1999). This is confusing, to say the least. In an effort to facilitate a standard definition of metadata, the National Information Standards Organization (NISO) defines it as "structured information that describes, explains, locates, or otherwise makes it easier to retrieve, use, or manage an information resource" (2004). In short, metadata allows machines to understand information.

This machine understanding of data is made capable through decisions by humans. In particular, humans decide how to parse out different elements of data to represent information. The primary focus of this chapter will be descriptive metadata—the scaffolding that provides details to represent information objects and help locate GI. However, there are other types of metadata that exist for purposes beyond

W. Bishop, T.H. Grubesic, *Geographic Information*, Springer Geography,
DOI 10.1007/978-3-319-22789-4_5

retrieval, including: administrative metadata (e.g., managerial information); preservation metadata (e.g., documenting actions taken related to versions); technical or structural metadata (e.g., software documentation or digitization compression ratios); and use metadata (e.g., tracking records or search logs) (Gilliland, 2008). We leave these data (and foci) to others (Greenberg, 2001; Medeiros, Bills, Blatchley, Pascale, & Weir, 2011).

Although metadata are often critical for data discoverability and reuse, in many instances, metadata might not be relevant. Context is important and metadata are relevant (or not) depending on their applicability to users and underlying purposes. Consider the wide range of metadata that consumers are exposed to during a single trip to a grocery store. Metadata on a customer's receipt at the grocery store represent the time of transaction, items purchased, their prices, and the total cost of the order. Within the store itself, metadata on the Food and Drug Administration (FDA) "nutrition fact labels" for each product detail serving sizes, calories per serving, and nutritional breakdown of the item (e.g., sugar, fat, carbohydrates, and so forth). Metadata in a grocery store's inventory database indicates the holdings and locations of items for sale (Hitt, 1996). In short, these types of metadata exist everywhere and manifest in many different and unexpected ways. At their base, metadata are truly data about data. However, the relative quality of these descriptive efforts help determine the findability of data for future use.

In revisiting the grocery store example, there are many nuances associated with the metadata that may not be readily apparent. For instance, although metadata contained within a customer's receipt quantify costs, they may also serve as a cue pertaining to consumer behavior (e.g., which products are frequently purchased in combination). Users of this monetary metadata also do not typically require an understanding of amortization or depreciation of value to purchase items they want. In fact, customers may not keep any version of their receipts beyond the purchase. Still, these metadata are important to store managers and the home office, where the aggregated data of customer purchases and their shopping habits will be put to good use. Other metadata, including information within the inventory database has uses related to managerial accounting (e.g., keep popular items in stock at cost-effective quantities for profit) and regulatory compliance (e.g., product recalls). No matter what the specific metadata consist of, users of these data typically have some type of specialized training or background that helps organizations measure, process, and communicate behavioral, financial, and logistical information using metadata. The key factor in financial information and its associated metadata is value, but for geographic information (GI), the key factor is location.

Metadata are pervasive and mostly simple in the physical world. Knowledge work in measuring physical items and organizing items into classes is longstanding and foundational to many hard sciences (e.g., Periodic Table). Information professionals working in libraries, museums, archives, and data centers have long dealt with often less tangible and more malleable information objects (e.g., documents). Information objects that require conceptual analysis to determine *aboutness* require different information organization approaches than the more concrete examples found in a grocery stores, chemistry labs, or rock collections. Subsequently, the knowledge organization and information representation required are more multifaceted. Metadata unique to GI, including information pertaining to scale, projection, and coordinates are necessary metadata to inform GI uses and users. Still, information representation of other important elements not inherently geographic are vital to the use of geospatial data and cartographic resources. While organizing surrogate records of information objects, users can *find*, *identify*, *select*, and *obtain* GI using the geographic and non-geographic elements.

These four uses (find, identify, select, and obtain) inform the extraction of specific metadata from information objects. Furthermore, these elements inform how to attribute the metadata to information objects, as well as who attributes the metadata to some extent (Taylor & Joudrey, 2009). As detailed earlier in this book, a variety of GI and associated policy exists to inform the impetus and details of metadata creation. However, even the geospatial data and cartographic resources without a federal mandate greatly benefit from the inclusion of some key metadata elements. Despite the perceived value of all metadata, very little research has been conducted on what makes a "good" metadata record for geospatial data or cartographic resources. The wide variety of types, complex formats, and real-time/streaming nature of some GI complicate finding the best approach for its information representation.

Unlike other information objects, GI presents knowledge workers with a palimpsest of information representations. Geospatial data (e.g., a geographic basefile of city streets) and cartographic resources (e.g., a roadmap) are already representations of geographic reality. Geographic information abstracts one particular view of geographic reality and serves as a substitute, through data or maps, something which enables users to experience and analyze the realities beyond human limitations of time and space (e.g., bird's-eye view) (Wood, 2010). As a result, GI does not always need additional metadata for use because creators and subsequent users have the data in front of them, on a map, and with a glance can access information pertaining to scale, projection, key attributes, and the like.

This general idea, rightly or wrongly, has been echoed in the geography and the geographic information system (GIS) communities for years. For example, in a personal communication with Michael Goodchild (2013), he suggested that creators and expert GI users will instantly know the metadata elements by simply looking at the data. In fact, this view anecdotally represents many GIS users' opinions, where statements such as "the only metadata *I* need is the phone number of the person who created the data" (Goodchild, 2013). While this may work for a small number of power users, the majority of GI users will lack the background and social network to simply place a phone call to determine the particulars of a database or any other type of GI. In fact, most users will require detailed metadata to assist them in finding, identifying, selecting, and obtaining GI. Further, there is power and intrinsic value embedded within metadata that can be leveraged for analytical purposes, enhancing discoverability, and the like. In the next section, we explore the value of metadata, especially those which helps users find, identify, select, and obtain GI. No phone numbers are included.

5.1.1 Metadata Value

It is no secret that both users and the uses of metadata vary widely. Further, the standardization of metadata, and all data for that matter, presents unique challenges. Again, consider the FDA nutrition fact label, which presents metadata that is standardized, and has value, despite disputes on the effectiveness of calorie labeling leading to decreases in overall consumption (Kiszko, Martinez, Abrams, & Elbel, 2014; Sinclair, Cooper, & Mansfield, 2014). For instance, in one study, only 37% of patients could calculate the number of carbohydrates consumed from a 20-ounce bottle of soda that contained 2.5 servings (Rothman et al., 2006). Acknowledging that metadata value only applies to those users with the literacies to access and use it may go without saying. Since most GI use requires some specialized training and tools the challenge for GI metadata is not in getting GI users to understand it, but convincing GI creators the value of metadata substantiates the investment in creating it.

Clearly, GI that lack metadata reduce their discoverability and if found GI users may not be able to assess the proper fitness for use. This results in less chance of access, use, or reuse for GI missing accompanying metadata and reduces that GI's value beyond its original creator. Information organization precipitates access and use. However, it is the latter (access and use) that are more easily (and frequently) counted metrics for evaluating data. This is troubling on several fronts. Consider, for example, the impact of wildfires on properties and homeowners. There are *real* property costs associated with fires, and prevention and suppression campaigns that facilitate a decline in the frequency and damage done by wildfires are justifiable. Such efforts help both individuals and communities to avoid much larger costs related to insurance and building efforts (Stephens, 2005). Obviously, few people would choose to have their property burn down, forcing a rebuild, rather than funding some type of fire prevention efforts. However, these types of shortsightedness problems manifest with data management all the time. In effect, any GI creator that would rather recreate and rebuild the data, repeatedly, instead of generating appropriate metadata that adheres to basic file naming conventions and helps assist in the relocation of GI, is drastically undervaluing metadata.

By using GI that helps inform wildfire prevention and suppression as an example, datasets concerning wildfire locations, population density, and distances from populated places and roads, provide a glimpse at discoverability issues. The descriptive metadata of these GI determine how they may be found, and metrics on access and use can be tallied (e.g., downloads). The search terms used (i.e., wildfires) may lead users to locate sources (e.g., U.S. Department of Agriculture, Forest Service nationwide spatial wildfire occurrence data) (Short, 2013). How users search for GI can inform metadata schema, but the value of that metadata remains implausible to calculate at the meta-scale because analytics for what is not found is never measured. Still, taking the time to create metadata is one metric and justifying the expense of information professionals to create metadata is a key question to address for metadata value.

Although a ton of work has been done on GI metadata, very little research has been conducted (or published). Specifically, metadata evangelists list many compelling reasons to create GI metadata and promote the inherent values of metadata

Table 5.1 Value reasons for metadata and potential measures

Value reason	Potential measure(s) of value and costs
Avoid duplication and protect the initial investment	GI creation, recreation, or purchase costs
Publicize and support	Web-analytic values (e.g., hit, page view, visit, and so forth)
Reduce workload related to answering questions	Information service costs (e.g., total question counts, average question negotiation times, and wages of information professionals)
Create institutional memory and data provenance	Data loss costs (e.g., accident, retirement, death, or other departure of knowledge worker) Data value of a retraceable, transparent workflow
Limit liability	Risk assessment of potential litigation

must have to justify its creation, but much more research needs to be conducted to verify and quantify this metadata value faith (Federal Geographic Data Committee, 2004; Martinolich, 2015). Evaluation of all services and resources in information agencies is vital to sustainable operations. Thus, the following sections present metadata value reasons with a discussion of several potential value and costs measures outlined for convenience in Table 5.1.

For return on investment (ROI) calculations, all values must subtract the variable costs of metadata creation. Metadata is an insurance plan for GI with the small deductible of the cost of creation. The labor cost to create metadata must also be computed as a key piece of all GI creation to avoid the potential sunk costs resulting from duplication, data loss, lack of sharing, distribution, or reuse. Information professionals' jobs that entail GI metadata creation, as well as cartographic resource description are calculable, but the costs depend on the institution, credentials, and other situational factors of the professional. The metadata value in these instances is *at least* commensurate with the benefits associated with using the data. If locating the data through a simple search engine query (and download) is possible because of adequate description, then no other means of information organization is required. Still, many examples of GI present more complex information retrieval issues than driving directions or determining the chance of precipitation for a city. These complexities can be accounted for, but the process must involve information professionals trained to create metadata and make the associated, specialized GI findable.

5.1.1.1 Avoid Duplication and Protect the Initial Investment

As detailed earlier, value quantification for metadata are typically associated with avoiding GI duplication, protecting investments, sharing, distributing, and reusing GI to save others the cost of recreating it. In these instances, where a lack metadata parallels to a total absence of the data, then the value of metadata is *equal* to the entire cost of GI creation. Specifically, if researchers, scientists, geographic information professionals, and other users are unable to locate GI, they may explore costly options related to purchasing or re-collecting the required data. As outlined in previous chapters, the creation and storage of GI may involve cost prohibitive activities such as sending surveying equipment into mangroves or mounting and launching remote sensing equipment on satellites into space. Further, for many local and regional governments, if it is possible to launch a single, concerted effort to collect a particular geographic basefile (e.g., street network), then distribute the collected data to other agencies and departments with a need for it, the presence of metadata for enhancing the ease of reuse and distribution is critical because it saves the costs associated with duplicative or re-creative efforts. It is important to remember that metadata value does not simply equate to its creation alone—but also because its standardization enables sharing and distribution.

In the not too distant past GI professionals and users when sharing large datasets had to exchange hard drives at conferences, or mail hard drives to collaborators. Today, the Internet and related cloud storage services (e.g., Dropbox) help to remove some of these technological barriers for sharing. However, one challenge that remains, hinges on metadata workflows and metadata that facilitate discovery (Durante & Hardy, 2015). The reduction of data duplication was a central aspect of U.S. GI policy discussed in the last chapter. Furthermore, newer federal policy directing that all federally funded research share their data (for advancing science), helps to inform decision-making as it relates to increased sharing and distribution (Fary & Owen, 2013). Similarly, some GI may only be collected at a single moment in time (e.g., climatological data). As a result, metadata are once again crucial for protecting the initial investment data collection—and at least as valuable as the GI itself. Again, GI must be organized according to certain values to be discoverable.

5.1.1.2 Publicize and Support

These metadata value claims do not necessarily equate to total GI loss, they do have some measurable worth that is frequently overlooked by the original GI and/or metadata creators. Specifically, metadata can help publicize and support GI by enabling users' queries to provide results. The action of GI being found, even if not used, provides value to the metadata. For example, discovering GI that has similar attributes to what is required for a project, even if it is not used, helps a user learn where to look and what to look for next. It also "plants a seed" for future projects by helping users understand what types of alternative data may be available. This suite of factors is relatively esoteric, but given the countless instances in which the publicity or support of some GI through its metadata may prove helpful does have value.

5.1.1.3 Reduce Workload Related to Answering Questions

There is value in metadata that helps answer questions for potential users. For example, depending upon the agency, data creators may not have the time or resources to respond to user queries about the GI contained within specific collections. There is a cost associated with such efforts, including the time required to phone, email, or chat about the data with potential users. Good metadata can answer these questions. Some models exist to calculate the actual costs of answering each question but to do so requires known values of the total question count, average question negotiation times, and wages. This type of evaluation may not be easily done outside of environments where timestamps and other transaction data are automatically captured (Bishop & Torrence, 2007). To conduct research on metadata value as it relates to the reduction of questions requires systematic tallying of all the times a user contacts the person or organization that created the data and the impetus for this may not exist in many information agencies.

5.1.1.4 Create Institutional Memory and Data Provenance

Oddly, these same values of publicity, support, and reduction in time dedicated to answering user questions are partially a result of creating institutional memory. A central theme in Information Science is the assumption that information will need to be found after its creation and in some cases in perpetuity. As Goodchild's comments on phoning the creator of the GI reflect a common pragmatic sentiment, if one considers the importance of long-term preservation, a number of issues arise where phoning does not suffice. Michener, Brunt, Helly, Kirchner, and Stafford (1997) provide a list of examples related to degradation of information over time that metadata may address: "general details about data collection are lost through time, accident may destroy data and documentation, retirement or career change makes access by scientists to 'mental storage' difficult or unlikely, and death of investigator and subsequent loss of remaining records" (Michener et al., 1997, p. 332). Institutional memory held in metadata ideally allows for GI immortality beyond any one postdoc on their way to bigger and better things. Without metadata, there may not be a way to know where the GI are, what the GI are, how the GI were created, or any other processing done to the GI.

Data provenance, also known as data lineage or pedigree, are technical metadata that describes the derivation history of GI. Documentation of this history enables retraceable paths to original sources, which removes the need to reinvest in the duplication of work. Data provenance increases the value, validity, and trust of GI by bolstering information quality, improving efficiency and adding transparency to workflow processes that allow for exploration beyond original data collection purposes. This is critically important for GI because the nature of GI manipulation, processing, and visualization requires significant manipulation (e.g., aggregation) and the fusion of data from many different sources. These data stewardship issues may not be in the minds of current GI creators focused on current deliverables. Then again, stewardship may not apply to everything created. Since the original creator might not always be at the phone or accurately recall data provenance in detail, metadata beyond the phone number of the original data creator is essential.

5.1.1.5 Limit Liability

Along these same lines, administrative and use metadata that outlines access and use constraints limits the liability of creators and presents another potential value of metadata. For example, if future users violate intellectual property rights or improperly use GI to navigate (perhaps getting lost or injured), good metadata may limit liability. Estimating the value of metadata, relative to these potential (albeit unlikely) costs, require further research and the development of metrics to quantify these elements. In reality, such efforts may be too difficult and time consuming to be worthwhile. The intrinsic value of knowledge depends greatly on what that information is and the intended audience. For example, the average reader probably has little interest in National Flood Insurance Program (NFIP) data when compared to insurers or those that live in a flood plain. However, the metadata explaining liability embedded within the NFIP helps insulate the GI creators from litigation that (sadly) always occurs after a catastrophic flood.

5.1.1.6 Metadata Are Valuable

For generations of information professionals, the inherent values of all types of metadata that enable discovery of documents, books, images, and other information objects remain understudied. In part, this is because the value of *knowledge*, *access*, and *use* did not require as much documentation and justification in the past, when compared to the present. Further, the lack of economic approaches to the study of information value is especially odd given that information representation and knowledge organization are core concepts in Information Science (Szostak, 2014). In fact, the first formal LIS program in the world was called the School of Library Economy (started by Melville Dewey at Columbia College in 1884). Still, as this chapter suggests, metadata creation and its associated terminology presents plenty of challenges to information scientists without the complications of conducting cost-benefit analyses. Discoverable information may help create an informed populace or lead to a more democratized society, but ultimately it is up to individuals and organizations to decide on whether or not incurring the costs of metadata creation is worth the immeasurable values outlined above.

5.1.2 Metadata Creation

As noted previously, metadata serves as the scaffolding that allows users to find, identify, select, and obtain GI. Metadata are purposefully structured to answer these questions and accomplish these tasks for users. The 1998 International Federation of Library Associations and Institutions (IFLA) *Functional Requirements for Bibliographic Records* (FRBR) repurposed these tasks for GI by establishing the ways relationships are represented between information objects. Some obvious differences in form exist when discovering GI and text-based documents, but the purpose of structuring relationships and the process of metadata creation remains strikingly similar. Thus, regardless of information object, users *may* or *may not* know what they are looking for, and metadata must address both user types. A key assumption of information organization is that users are actually trying to find *something*. Some information scientists study more complex aspects of human information seeking behavior like serendipity and imposed queries (Makri, 2014), but all formal information organization sets out to connect people with the information they purposefully seek in a structured manner.

Given this context, it is important to revisit the core, enabling elements of metadata: *find*, *identify*, *select*, and *obtain*. Although these general elements were initially structured to help users find bibliographic material (Cutter, Ford, Phillips, & Conneck, 1904), it is their generality that allows them to be applied to GI over 100 years later. Specifically, metadata were (1) "to enable a person to find a book of which either the author, title, or subject were known"; and (2) "to show what a library has by a given author, on a given subject, or in a given kind of literature" (p. 12). IFLA built upon Cutter's (Cutter et al., 1904) purpose, as well as other advancements in the various information professions, to modernize the framework for constructing high-quality, descriptive metadata. Where GI is concerned, *find* simply helps users discover GI using search criteria that relates to attributes of an entity. In other words, basic information retrieval actions allow for related GI to appear during a search. *Identify* allows a user to confirm that they have found what they were searching for. More importantly, *identify* also allows users to distinguish GI from similar resources. For example, when a Central Intelligence Agency (CIA) published map of Syria is found, users must have the ability to distinguish between versions and publication dates. *Select* is a metadata function that provides users with the needed information to determine if the GI meets their project needs, with respect to content, physical format, and so forth. For GI, a resource may have been collected in the wrong datum or at an unusable scale and users need this information to determine fitness for use. *Obtain* enables users to retrieve GI. This may be as simple as downloading a compressed file, or as complicated as going to a map library and pulling a sheet from a drawer. In most GI, it may be difficult to determine where the data ends and the metadata begins. Print cartographic resources and many aerial photographs contain descriptive metadata within the information objects themselves (e.g., text, basic cartographic guidelines, and so forth) (Danko, 2012). Regardless, access and use can only occur when a user's discovery needs (i.e., find, identify, select, and obtain) are met. Needless to say, this requires organizations and data creators to anticipate those needs and do their best to make the data both discoverable and well described. Table 5.2 details the core elements of high-quality, descriptive metadata. The examples are drawn from the Dublin Core (DC) Metadata Element Set versions 1.1. The DC is the result of collaborative work by an international, cross-disciplinary collective in librarianship, computer science, text encoding, museums, and other fields participating in the Dublin Core Metadata Initiative (DCMI) (2015)—with a goal of creating broad level and core elements for use across fields.

Table 5.2 Questions metadata answers

User questions	Metadata elements
Who …created/processed the GI? …wrote the metadata? …can I contact with questions? …owns the GI? …is responsible for the GI?	Creator Contributor Publisher Rights
What …are the scale, projection, coordinate system of the GI? …study did the GI come from? …file format is it? …constraints of use? …appropriate uses? …resources are related? …source did the GI derive from? …is the GI about? …what is the nature or genre of the GI?	Coverage Description Format Language Relation Rights Source Subject Title Type
Where …were the GI created? …is the spatial applicability of the resources? …is the jurisdiction under which the resource is relevant? …is the GI located now?	Coverage Identifier
When …were the GI created? …were the metadata created?	Date
Why …were the GI created?	Description Relation Source Subject
How …were the GI created? …do I get the GI? …much does it cost?	Coverage Description

5.1.3 A Conceptual Model for GI Representation

IFLA's FRBR also resulted in a framework to assist in understanding the terms and functions of information representation to meet user needs. Despite a focus on bibliographic description, the FRBR elements provide a useful approach for discussing any information object and its surrogate records. A surrogate record is an abstraction of the information about a document that assists users in finding, identifying, selecting, and obtaining that information. Again, a surrogate record is an Information Science term for the metadata files that are not the information object themselves. From the century of the card catalog (1870s–1970s) to the dawn of Online Public Access Catalog (OPAC), the record and information object were clearly separate entities, but in a digital and networked environment the distinction is less clear. Geographic information and corresponding metadata may live on different servers, but the end-user does not perceive this because most information systems are structured to operate seamlessly during the retrieval process.

The FRBR framework was based on an early entity-relationship model (Martin, 1982). Interestingly, the Resource Description Framework (RDF) developed by the World Wide Web Consortium operates in the same manner with statements that take the form of subject-predicate-object (i.e., triples in ontology literature). Much like the RDF, the FRBR consists of three elements, *entity-relationship-attribute*. Specifically, within a domain, a collection of objects, called *entities*, has *relationships* among those objects and those entities have particular *attributes*. For example, a GI entity, such as Topologically Integrated Geographic Encoding and Referencing (TIGER)/Line file, would have many relationships with other entities, like its creator the U.S. Census Bureau, including its coverage of the political boundaries, and its subject of socio-demographic data. Building the relationships in a structured way contextualizes the links between sources. This enables more complex queries to be performed across data and provides structure to the relationships between information objects.

Drilling deeper, there are three groups of entities that help describe an information object. Group 1 entities represent information objects of intellectual or artistic creativity. TIGER/Line data are an example of a Group 1 entity, but this would also include any map or cartographic resource. Group 2 entities represent who is responsible for the information object (e.g., U.S. Census Bureau). Any geographic information professional or organization that create GI can be categorized as a Group 2 entity. Group 3 entities include the subjects related to the information objects. For example, the socio-demographic data that can be connected via TIGER/Line administrative boundaries would qualify. That said, Group 3 entities present more of a contextual challenge, because arranging subjects and ascribing them to information objects can be difficult, especially when compared to Group 1 and 2 entities—both of which are very concrete in GI.

For what it is worth, FRBR also introduced a Work, Expression, Manifestation, Item (WEMI) model to disambiguate Group 1 entities. On the face of it, a map is a map, and a file is a file. Thus, when working with personal or small collections the WEMI model is not practical. For larger, real-world collections, few one-to-one connections between works and items exist. As a result, providing a structure to those relationships helps users distinguish between items that are similar, but not the same. Consider Fig. 5.1, which is a publicly available map from the Census Bureau (http://tinyurl.com/h6mcw28) that represents the change in total population by state between 1900 and 2000. Here, Fig. 5.1 is a *work*, realized through an *expression* that is embodied in a *manifestation* that is exemplified by an *item*. Although practicing information professionals criticize the model because it lacks utility, WEMI does present a useful distinction to address versioning issues of all information objects, including maps like Fig. 5.1.

5.1.3.1 Work

To be sure, the concept of a *work* is very abstract, but is defined in FRBR as a distinct, intellectual, or artistic creation. IFLA acknowledges that different cultures or national groups will have various interpretations of boundary distinction. That said, a TIGER/

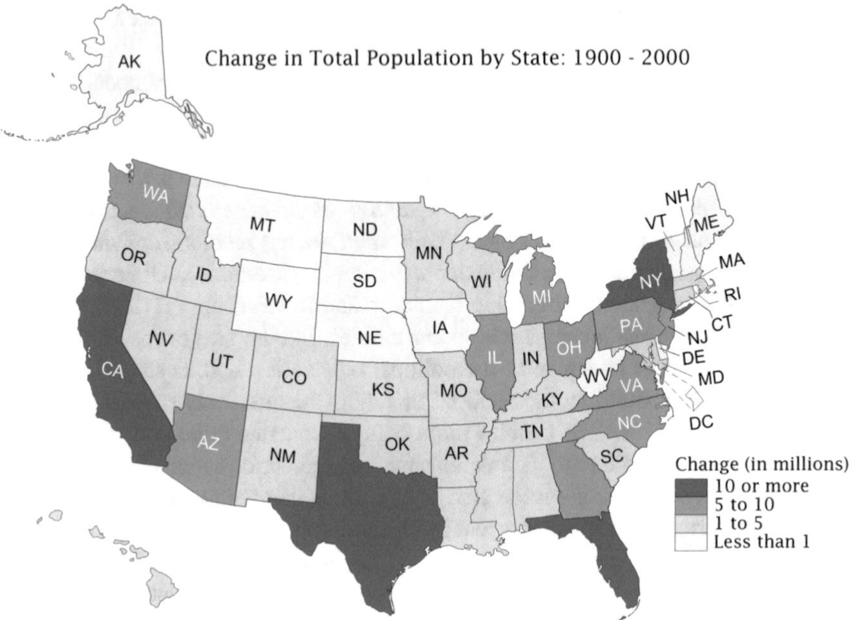

Fig. 5.1 Change in Total Population by State: 1900–2000

Line file is representative of a work. So is the 1482 edition of Ptolemy's world map printed in Ulm, Germany. It is also important to note that significant modification of a work results in another work. For example, the 2015 TIGER/Line and the 2014 TIGER/Line files are different works, but the Ptolemy map hanging in a conference room and the image in your browser share the same root of an idea and are the same work. For GI, there are several attributes that make a work in this domain, unique—including geographic coordinates and equinox. Specifically, these work attributes adhere the GI onto a specific geographic or celestial reference to define specific parts of the world that are being represented (e.g., the U.S., Germany, and so forth). IFLA disambiguates works by citing William Shakespeare's play *Romeo and Juliet* and the films based on the play. The films are significant modifications and thus new works. In text-based information objects, there is a myriad of paraphrasing, rewriting, adaptations, parodies, and these variations result in new works.

One may imagine that cartographers would have a field-day when attempting to delineate works. For example, if students in a cartography class were given an identical set of data and asked to produce a map, the students would likely produce unique works. To quote Monmonier (2008, p. 2), "a single map is but one of an indefinitely large number of maps that might be produced for the same situation or from the same data". That is to say, the students' thoughts about how to interpret the task would be different and so too would be the resulting works. In this example, although the geospatial data the students used to start their task would be the same work, the resulting maps would not be.

One challenging aspect to the FRBR is developing an understanding that work does not exist outside the creator's mind until it is realized as an expression. For example, the cartography class knows they will make a map of the U.S. and each student will produce their work with the same data. However, even though defining intellectual artifacts inside the mind does not assist in organization, access, or use, the acknowledgment that information objects start as ideas does form the basis of the WEMI model to distinguish those objects and justifies intellectual property claims.

5.1.3.2 Expression

An expression is the form a work takes when it is realized. For books, the expression is the specific words, sentences, paragraphs, that create a unique expression of a work. Subsequently, any change in form results in a new expression. Expression attributes that relate to GI include scale and projection (there are others). The GI in a TIGER/line file could be expressed in any number of ways in a GIS, depending on the purpose of the analysis and geovisualization. For example, the expressions generated by students in the cartography class may vary depending on each student's choice of cartographic elements. However, even if *every* line thickness was the same and each map element were positioned exactly the same by the same student, the map would be a different expression of the same work. For the Ptolemy world map, each online image, each variation in color, each printing (e.g., from the 1482 edition of *Cosmographia*, to those printed on wall posters and pillows) represent different expressions of the same work. It is important to remember, however, that an expression is still an abstract concept. It is but one realization, yet it is still not a tangible information object that takes a physical form.

5.1.3.3 Manifestation

A manifestation is the physical embodiment of an expression. Each reprojection of the same data creates a new expression and a new manifestation. For example, scanning a hardcopy cartographic resource at different resolutions would create different manifestations of the same expression. The TIGER/Line file manifestation occurs the moment the data are real, complete, and posted on the Census website for download. Again, each manifestation of the maps created in the cartography class to represent demographic change is a new manifestation. If a student prints the map and creates an image of the same map, each map is a different manifestation of the same expression. For whatever reason, the Census Bureau map displayed in Fig. 5.1 has several state abbreviations missing. At some point, perhaps the Census Bureau will catch these errors and create another manifestation of the same map expression.

For Ptolemy, if the same scan of an expression of the world map from the Ulm publication is printed on a postcard, or appears online, each is a different manifestation of the same expression. The distinctions may seem trivial for most purposes, but in data provenance and information representation each difference is meaningful, at least to some degree. Consider, for example, GI from the Geoweb. Resources are continually being updated and this creates an infinite number of manifestations of the same expression. These minute distinctions only matter when a particular manifestation is what the user seeks. With 78% of the records in WorldCat having only one manifestation the distinctions do not have much practical use for any information professional dealing with bibliographic resources (Bennett, Lavoie, & O'Neill, 2003). However, "cartographic resources rarely have revisions that would *not* be considered a new works, expressions, and manifestations" (Andrew, Moore, & Larsgaard, 2015, p. 14).

5.1.3.4 Item

The final piece of the WEMI model is the item. An item is the single exemplar of a manifestation. For GI, this is the single file or print map. These are the actual entities that need to be described with metadata—the items. In short, this means that the TIGER/Line file on any given computer is a different item than the TIGER/Line file on any other computer. In giving thought to how all works, expressions, manifestations, and items relate, it can assist information professionals in determining what metadata are useful. It also proves useful in distinguishing different information objects from each other and establishing a relationship between like items.

In sum, FRBR provides a thoughtful account of what metadata to address and a conceptual framework to distinguish between different facets of information objects. However, readers know that this is not a simple task and many challenges remain. Fortunately, the domains of knowledge organization and information representation have a much longer history in addressing such issues. Moreover, the terminology and approaches for GI organization are still being developed, despite the vastly different and/or special nature of GI.

5.2 Knowledge Organization Concepts and Tools

Knowledge organization is critical to an understanding of current and future metadata. Each operational choice, whether it be technology, standard, and/or schema influences how future users will be able to find, identify, select, and obtain GI. These topics do not require rediscovery, reinvention, or reimagining across every discipline for at least two reasons; (1) viral replication of approaches to metadata complicate data discovery and sharing information across domains; and (2) other realms of information management would save time by building off of Information Science's intellectual contributions and experience. Information Science is unique in that its longstanding tradition is to have a multidisciplinary focus and manage information across domains.

Scholarly communication itself has benefitted directly from advances in indexing, knowledge organization, and information retrieval by making text-based discovery of scholarship both seamless and efficient. There is no doubt that the creation of Science Citation Index (SCI) by Eugene Garfield in 1964 impacted anyone working in higher education. SCI (and related tools) fueled the growth of Scientometrics, as an important subfield. Co-citation analysis and its many variations provides scholars with a deeper understanding of knowledge domains, collaborative scientific networks, and scientific impact. More importantly, these types of deep-dives into scientific collaboration and impact are only possible because of the metadata structures underlying all indexed scholarly communication—which was started by documentalists.

To be sure, disciplines other than Information Science have made important contributions to these efforts. However, taking a moment to look back at the last 140 years of knowledge work gives some terminology and context to help readers fully appreciate the clockwork of metadata. There is no need to start from the beginning, but it is worth revisiting the basics. For example, the seemingly simple principles of authority control, vocabulary control, classification, and ontologies become quite complex when dealing with information beyond the scale of individual collections or projects.

5.2.1 Authority Control and Vocabulary Control

As detailed above, Group 2 and Group 3 entities present unique challenges to information organization with many different synonymous concepts, word form flexibility, homographs, homophones, abbreviations, and acronyms. For Information Science, the concept of authority control emerged to manage the challenges associated with creators that shared the same name, or when names changed over time. The ability to determine the identity of a creator and distinguish that person from others with the same name is extremely useful. Further, if one can combine all the works by the same creator that used different names over time, it is possible to locate nearly everything created by one person. For example, in the Library of Congress Authorities, the authorized heading Twain, Mark, 1835–1910 allows users to find all works by Samuel Langhorne Clemens, Louis de Conte, and Quintus Curtius Snodgrass. Although this type of excessive pseudonym use is more of an exception, rather than a norm, especially for GI creation, individuals do change their names for various reasons (e.g., marriage, divorce, and so forth). The Library of Congress Authorities has a file with value added information to explain name changes and attributes about individuals. This knowledge organization tool could be repurposed, as it is quite exhaustive, but the idea of authority control and how it overcomes some information organization problems is important for informing GI metadata creation.

The English language proves to be especially challenging when it comes to vocabulary control. Like authority control, a controlled vocabulary allows users to find every information object about a particular subject. Vocabulary can be specific (e.g., Maine Coon) or general (e.g., cats). More importantly, the meaning of a word or term can change when users alter the sequence and form of multiword terms. In order to coordinate terms that are related, link synonyms, disambiguate hypernyms, hyponyms, homonyms, has developed many controlled vocabularies and thesauri that the field of

Information Science enable users to find all information on the same subject within indexed collections. The Library of Congress Subject Headings (LCSH), Medical Subject Headings (MeSH), Art & Architecture Thesaurus (AAT), and Educational Resources Information Center (ERIC) are some of the most used and useful tools for controlling vocabulary. In the age of folksonomies, where humans hashtag their way through the day, controlling vocabulary might appear antiquated. Still, this type of Group 3 metadata is very useful and allows for more precise and accurate information retrieval in information systems. For example, #awesome does not make GI more findable because all GI are awesome and both novice and expert users will likely be searching for a more unambiguous topic when seeking information.

Finally, gazetteers are a very specific type of controlled vocabulary developed in concert between geographers and information scientists. A gazetteer is an authoritative registry of placenames, often with spatial footprints and feature descriptions (Buchel & Hill, 2011). In discovery use cases, users wish to search and browse by location, and gazetteers enable the categorical use of placenames relative to geospatial location. Increasingly gazetteer services, such as the open source, community-editable GeoNames, provide linked data interfaces into the placename registry. These interfaces enable new types of discovery uses, such as hierarchical spatial browsing, similarity search, and ontological queries (Kessler, Janowicz, & Bishr, 2009). For GI metadata, placename standardization is central to the organization despite spatial ontologies allowing for crosswalks across languages and time (Bishop, Moulaison, & Burwell, 2015). Gazetteers enable users to find everything related to a particular place. These authority, vocabulary, and place-name control tools were built over time and are well-tested. Thus, there is no need to reinvent the wheel for this type of control structure and several linked data initiatives in information agencies parlay these older structures to facilitate semantic searches (Byrne & Goddard, 2010).

5.2.2 *Classification Systems*

In 1876, Melville Dewey published the Dewey Decimal Classification (DDC) System with the intent of organizing all knowledge into classes and grouping like-items together to make retrieval easier. The system was hierarchical, much like Carl Linnaeus's taxonomy (Linnaeus, 1758) of biology, where each species has a place in kingdom, phylum, class, order, family, and genus. The DDC organized all knowledge into ten classes, each divided into ten divisions, each having ten sections, and within each section allotting for further division by tens infinitum. This enumerative nature of the decimals allowed each subject to have a designation. Another advantage to using decimals is that they are infinitely expansive; however, the hierarchical structure of the DDC presented fundamental problems for information objects that could belong to more than one class. The system also retained biases in that the developed classes devote more space to cultures, geographies—reflecting the ideals of its creator. Regardless, the DDC remains the most used classification system in at least 135 countries (Online Computer Library Center, 2015).

With georeferencing, GI has built in literal colocation in information organization; however, the use of a classification system to arrange subjects that are not bound by space in a logical manner often leads to the discovery of alike items. Thus, true serendipity does not occur in data discovery because metadata requires that a certain structure of knowledge organization is built into any collection of objects. Where GI is concerned, many user are directed to finding *known* GI. However, knowledge organization permits these queries to uncover GI that may not have been known or anticipated by the user. This is especially important for data discovery because many users are asking questions as a result of a knowledge gap. This means they are searching for information with an original, often-ambiguous query (Taylor, 1968). In short, some elements of knowledge organization facilitates discovery, via search, even when users do not know how to look, or what they are looking for.

To overcome the limitations of hierarchical design several faceted classification systems emerged, such as the Universal Decimal Classification in 1892 and Colon Classification in 1933. Faceted classifications allow for some categories to be categorized across classes and adds another dimension to the search. For example, the Colon Classification had five fundamental categories to use as facets across subjects; personality, material, energy, space, and time (Taylor & Joudrey, 2009). Clearly, space and time are key elements across all GI even if those charged with managing GI never conceptualized those metadata as facets.

While classification systems provide a means of organizing knowledge into categories, ontologies arrange a domain of knowledge based on the structure of objects in a domain and their relations (i.e., facets). The visions of linked data and the Semantic web require the building blocks of ontologies to function. With the relationships of information objects in a knowledge domain formally articulated, new types of automated reasoning can occur. For example, connections unknown to a user could be presented and help to reveal new information about related concepts, automatically. Group 3 entities of subjects rely upon classification from specific fields (e.g., geology), but some crosswalks between different classifications are still necessary to determine fitness for use with GI. Geographic information has various information types, complex data formats, ancillary files, and exhibits dynamism. Further, it is voluminous, some of the data are proprietary and most require domain-specific metadata for appraisal and subsequent use. To organize the entire corpus of GI, classification is necessary. But, knowing the strengths and weaknesses of enumerative, expansive, and faceted classification systems already developed should guide future classification construction. Many of the schemas, profiles, and standards discussed next build off of the concepts of authority control, vocabulary control, and classification.

5.3 Schemas, Profiles, and Standards

The purpose of this section is to present some actual examples of digital standards, profiles, and cartographic resource descriptions. Having covered the definitions, value, purpose, and some knowledge organization tools previously, readers should be prepared to explore some of the detailed elements of actual metadata. Unfortunately, too often the background that allows users to find, identify, select, and obtain GI is presented superficially. The purpose of this chapter was to provide a concise overview to demonstrate how metadata creation are not always common sense tasks and that this complexity has been an old concern for information scientists. Further, although the focus of this chapter is descriptive metadata, the importance of administrative metadata (e.g., managerial information) and structural metadata (e.g., technical information) should not be forgotten. Both remain critical for information management throughout the data lifecycle.

The Dublin Core (DC) metadata standard is a set of metadata elements designed for a specific purpose. Schemas are designed to reflect a certain set of instructions for the metadata, including the elements, semantics, and content rules. Although tedious, these rules are important because they dictate punctuation, capitalization, and other rules on allowable values in metadata fields. Over time, standards evolve from a schema when a community adopts a particular set of elements, semantics, and rules. In turn, this cauterizes them as published and with enforceable compliance. For example, Darwin Core (DwC) provides one example of a standard that is an offshoot from the DC. DwC uses the same schema, but with a particular focus on biodiversity informatics with extensions related to taxonomy, location, and geology at a higher granularity than required for something bibliographic. Bibliographic works may have many related iterations, editions, and printings, but a field like biology requires greater specifics to differentiate within species. Each community will adopt their own standards and due to frequently changing technology, GI geospatial metadata are not as rigorously standardized as those for many other information objects. The following sections present one digital and one mostly hardcopy standard.

5.3.1 *One Standard and One Profile*

As outline in Chap. 4, the Federal Geographic Data Committee (FGDC) was formed to promote the "coordinated development, use, sharing, and dissemination of geospatial data on a national basis" (2015a). The FGDC has many charges related to the implementation of the National Spatial Data Infrastructure (NSDI). In June 1998, the FGDC approved a metadata standard Content Standard for Digital Geospatial Metadata (CSDGM) to provide a common set of terminology and definitions for the documentation of digital geospatial data. At present, the FGDC encourages the use standards ISO 19100:2003 Geographic information—Metadata and ISO 19115-2:2009 Geographic information—Metadata—Part 2. Implementation is ongoing with XML schemas on GitHub. Still, most GI retains this particular legacy metadata standard. Once one standard is learned, crosswalking to other standards is relatively easy both cerebrally and literally with transformations.

The Content Standard for Digital Geospatial Metadata Workbook (Federal Geographic Data Committee, 2000) has seven main sections and three supporting sections, all designed to outline how elements are grouped, which are mandatory (or not), as well as those that can repeat. Readers be warned, the CSDGM can be difficult to read as graphical representation was used to imbue meaning. For example, compound elements were depicted using a simple two-dimensional box and data elements were depicted using a three-dimensional box. Also, there was a color coding scheme used with yellow indicating mandatory data elements. If one gets past this overly complex presentation, a clear standard with key elements emerges and the workbook outlines in great detail the type of values applicable for each element. The seven main sections include: (1) Identification Information; (2) Data Quality Information; (3) Spatial Organization Information; (4) Spatial Reference Information; (5) Entity and Attribute

Table 5.3 CSDGM elements and example

Identification Information Element	Brief examples
Originator	<origin>George Aiken Dave Krabbenhoft, Orem, Bill</origin>
Publication Date	<pubdate>2005</pubdate>
Title	<title>Interactions of Mercury with Dissolved Organic Carbon in the Florida Everglades</title>
Abstract	<abstract>This project is designed to more clearly define the factors that control the occurrence, nature, and reactivity of dissolved organic matter (DOM) in the Florida Everglades…</abstract>
Purpose	<purpose>This research is relevant because of the high natural production of organic carbon in the peat soils and wetlands…</purpose>
Time Period Information	<begdate>199503</begdate><enddate>2006</enddate>
Coordinates	<bounding> <westbc>-80.891142</westbc> <eastbc>-80.102985</eastbc> <northbc>26.78571</northbc> <southbc>25.597272</southbc></bounding>
Theme Keywords	<themekey>mercury</themekey> <themekey>dissolved organic carbon</themekey>
Access Constraints	<accconst>none</accconst>
Use Constraints	<useconst>None. Acknowledgement of the U.S. Geological Survey would be appreciated for products derived from these data.</useconst>.
Metadata Date	<metd>20070126</metd>
Contact Information	<cntperp><cntper>George Aiken</cntper> <cntorg>U.S. Geological Survey</cntorg></cntperp> <cntpos>Project chief</cntpos><cntaddr> <addrtype>mailing address</addrtype> <address>3215 Marine Street, Suite E-127</address> <city>Boulder</city><state>CO</state><postal>80303</postal></cntaddr> <cntvoice>303 541-3036</cntvoice> <cntfax>303 447-2505</cntfax> <cntemail>graiken@usgs.gov</cntemail></cntinfo>
Metadata standard and version	<metstdn>Content Standard for Digital Geospatial Metadata</metstdn> <metstdv>FGDC-STD-001-1998</metstdv>

Information; (6) Distribution Information; and (7) Metadata Reference Information. There also remain three supporting sections: (1) Citation Information; (2) Time Period Information; and (3) Contact Information. Table 5.3 presents a minimal metadata for a CSDGM record that is drawn from the U.S. Geological Survey (USGS) (2015). In this instance, only the Identification Information, Metadata Reference Information, Citation Information, Time Period Information, and Contact Information are presented—all of which are mandatory.

Readers should note that this is just one section of a complete record. It is also important to highlight that metadata creation takes significant time and because of its complexity, is prone to error. A recent review of 617 CSDGM records from twelve USGS metadata collections found several mandatory field missing, including publisher (40.8%); resource type (47.4%), and resource format (90.4%) (Kozimor, Habermann, Gordon, Powers, & Farley, 2016). These metadata creation errors reflect a propagation of errors coming from data itself. For example, one review of data papers in an ecological society archive found the majority of papers (92.5%) did not describe collection methods, data collection sites or time frames, and left out other relevant variables to enable reuse of the data and metadata creation (Kervin, Michener, & Cook, 2013). Sadly, this means that mandatory elements that help answer basic user questions to help find, identify, select, and obtain GI are likely missing from many sources.

Initially, the CSDGM was structured to precipitate several domain specific profiles that included extensions. A profile extends the base standard by adding metadata elements to meet their specific community metadata requirements. To date, only one profile, the Biological Data Profile, can be considered a success. Much like DwC, the Biological Profile added Taxonomy to the full classification. The Shoreline Profile was created in 2001, but the FGDC quickly moved on to the ISO standards and implementation is ongoing. Several other standards exist that may be applied to GI, including DC. The FGDC site has a good list of geospatial metadata tools (Federal Geographic Data Committee, 2015b) (http://www.fgdc.gov/metadata/geospatial-metadata-tools).

Great strides have occurred in the information professions recently, to reduce some of the challenges of working with multiple metadata standards operating simultaneously with much GI retaining legacy metadata. GeoBlacklight allows users to discover through a federated search across multi-institutional repositories of GI, but for this function to work developers had to create a schema that reconciles core metadata elements allowing for the same search across FGDC and ISO records (Hardy & Durante, 2014). The functions that allow for auto-fill of elements and batch-editing expedite metadata creation processes. The OpenGeoPortal Metadata Working Group (http://opengeoportal.org/working-groups/metadata/), a multi-institutional partnership of data and metadata experts, oversees the development of new tools and establishment of best practices for creating and exchanging GI metadata. For digital cartographic resources and other geospatial data, the metadata creation future is bright.

5.3.2 *Cataloging of Non-Digital (Hardcopy) Cartographic Resources*

In Dresden, near the end of the eighteenth century at the *Kurfurstliche Bibliothek,* exists the earliest noted practice of map cataloging (Klemp, 1982). Elsewhere, the earliest known formal map catalog (i.e., a list of all maps in a library) in the U.S. was produced at Harvard in 1831 (Drazniowsky, 1966). Since these early efforts, the cataloging fate of non-digital cartographic resources has been victim to economics and bibliographic control. As the foremost map librarian in the U.S., Mary Larsgaard (1998) speculates that librarians could not justify spending the same amount of time creating a surrogate record for a map as a book. This simplistic economic choice led generations of librarians to measure the "worth of printed work [...] by size and weight" (p. 158). This resulted in many maps existing without metadata to help users to find, identify, select, and obtain needed GI.

Despite this second-class treatment when compared to books, many maps have been cataloged using a series of codes from the American Library Association's and British Library Association's *Catalog rules: author and title entries* (American Library Association, 1908). As daunting as they sound, the Anglo-American Cataloguing Rules (ACCR) (1967) and Anglo-American Cataloguing Rules, Second Edition (ACCR2) (1978) held dominion over the process of cataloging of non-digital (hardcopy) cartographic resources for many years. With the roots of these rules in bibliographic control, some of the concepts did not translate well for describing cartographic resources in a findable way. For example, the authority and access point of the organization or corporation often take precedent over other elements that are more important to locating books, such as author and title. That is to say, a cartographer or title of a map might be irrelevant, secondary, or absent. Instead, more emphasis from the user's perspective is placed on the organization creating a map or the actual map subject. Also, the locality of an information object may be the most important element of a resource. These access issues of prior cataloging rules were not only impacting maps. As outlined previously, IFLA's FRBR provided guidance for future cataloging efforts, refocusing the purpose of cataloging more squarely on users' needs. The result was Resource Description and Access (RDA) in 2010 that makes information representation more human and machine readable.

Today, RDA allows for "the chief source of information for cartographic resources is the entire item itself" (Andrew et al., 2015). "Take what you see" permeates RDA instructions and since cartographic resources are designed for users, seeing all the information is powerful. In short, not much digging is required to locate needed information, other than turning a map over and checking the back for additional detail. Common cartographic practice is to write the entire map title in uppercase and when practicing "take what you see" catalogers will copy the title in the same manner and unlike AACR2 rules no longer alter to title case. Interested readers should turn to the *RDA and Cartographic Resources* (Andrew et al., 2015) for a more encyclopedic coverage. The changes from AACR2 get excessively detailed, quickly. For lay persons, it is important to acknowledge that RDA has slain almost all abbreviations and odd Latin phrases that only professional catalogers would understand. Oddly, making records more machine and human readable reduces keystrokes and the specialization required for catalogers to deal with non-digital cartographic resources.

Core elements in RDA remain similar to those outlined in other metadata schema for GI: title, creator, scale, and physical description. Still, a shadow of the errors introduced by bibliographic control remains in some RDA suggestions. For example, RDA suggests if a scale is not found to consult a similar resource to locate scale. Since scale indicates how data was collected and determines fitness for use, consulting a similar resource would be very imprecise, at the least, and likely dangerous. Experts encourage unknown scales to be indicated in the record as 'Scale not given' and reduce the chance of misleading users by providing a scale not derived from the item. To be clear, RDA is not a good fit for many GI types, including those with complex data formats, big GI, or that which requires domain-specific metadata for appraisal and use. In short, most GI, including digital cartographic manifestations, should use other schema.

For more on the topic of cataloging print cartographic resources consult Larsgaard and Andrew's co-edited *Maps and Related Cartographic Materials: Cataloging, Classification and Bibliographic Control* (Andrew & Larsgaard, 1999). Extensive cataloging training is also reviewed in all three editions of Larsgaard's classic *Map Librarianship: An introductio*n. Prior to RDA, Andrew created the still useful manual, *Cataloging Sheet Maps*: *The Basics* (Andrew, 2003). The *Journal of Map and Geography Libraries (JMGL)* and *Cataloging and Classification Quarterly* both contain many other significant contributions on this topic. The distinction between the roles of information professionals and others working with GI may be most easily seen in print. Catalogers do not need to know what any projection means, but only that it matters to users and to transcribe that into records. Much of the necessary metadata creation skills required to assist in geospatial data discovery will require deep knowledge of Information Science, and likely only some basic knowledge of Geography.

5.4 Conclusion

Because most GI represents an abstraction of reality, there are a number of inherent challenges that need to be dealt with to have GI make sense. For example, representing a round object on two-dimensions requires compromising some qualities of space (i.e., shape, area, distance, direction). Metadata are no different. In Lewis Carroll's (1893) *Sylvie and Bruno Concluded*, a character claims to make a useful map for the entire country on a scale of one mile to one mile (1:1). As authors, we hope that all the well-informed geographic information professionals reading this anecdote are "in" on this joke and recognize the absurdity of this concept. The farmers of the country would not let him open the map as a map at that scale would cover everything, block out the sun, and kill their crops. Developing some type of metadata for this map would be equally challenging.

Given the richness of the real-world's geography, representing particular attributes requires simplification and abstraction. All of these distortions are necessary to fit multidimensional geographic realities onto two-dimensional sheets or within digital environments. A great deal of GI was created and continues to be created in these two dimensions. Therefore, geographic elements noting accuracy and precision remain paramount for many users determination of the fitness for use for GI and its associated metadata. So, how much metadata is needed? It depends. Minimal metadata is minimally useful and too much metadata complicates GI access and use.

The list here is not comprehensive, but a few maxims that anyone should employ related to FRBR entities are:

- Avoid jargon (Group 3 entities);
- Define technical terms and acronyms (Group 3 entities);
- Complete titles include What, Where, When, Who and Scale (e.g., Greater Yellowstone Rivers from 1:126,700 Forest Visitor Maps (1961–1983));
- Use fully qualified geographic names from gazetteers (Group 3 entities);
- Use thesauri whenever possible for subject (Group 3 entities);
- Use authority files whenever possible for creators (Group 2 entities);
- Liability statements for access and use (Group 1 entities); and
- Anticipate potential users' information needs and behaviors.

Information Science provides the foundation for digging deeper into relevant conceptual frameworks and terminology—helping to steer the study of GI knowledge organization and information representation.

In total, Part I of the book provided a review of several key topics encompassing GI organization. A review of geographic representation types and core components like geographic scale, projections, and coordinate systems gave readers a baseline understanding of GI creation. GI policy, such as SDI, and metadata, its value, and purposes with discussions of other knowledge organization concepts and examples of geospatial metadata schemas, profiles, and standards, solidified all the essentials to comprehend GI organization. Part II of the book shifts toward a focus on the new platforms that enable geospatial data discovery to satisfy the growing number and variety of GI information needs related to GI access and use.

References

American Library Association (1908). *Catalog rules: Author and title entries. Comp. by committees of the American Library Association and the (British) Library Association*. Chicago: American Library Association Publishing Board.

Andrew, P. G. (2003). *Cataloging sheet maps: The basics*. New York: Haworth Information Press.

Andrew, P. G., & Larsgaard, M. L. (1999). *Maps and related cartographic materials: Cataloging, classification and bibliographic control*. Binghamton, NY: Haworth Information Press.

Andrew, P. G., Moore, S. M., & Larsgaard, M. L. (2015). *Resource description and access and cartographic resources*. Chicago: American Library Association.

Bennett, R., Lavoie, B. F., & O'Neill, E. T. (2003). The concept of a work in WorldCat: An application of FRBR. *Library Collections, Acquisitions, and Technical Services, 27*(1), 45–59. doi:10.1080/14649055.2003.10765895.

Bishop, B. W., & Torrence, M. (2007). Virtual reference services: Consortium vs. stand-alone. *College and Undergraduate Libraries, 13*(4), 117–127. doi:10.1300/J106v13n04_08.

Bishop, B. W., Moulaison, H. L., & Burwell, C. L. (2015). Geographic knowledge organization: Critical cartographic cataloging and place-names in the geoweb. *Knowledge Organization, 42*(4), 199–210.

Brady, P., Parker, T., Stone, M., (Writers), & Parker, T., (Director). (1999). Starvin' marvin [Television series episode]. In T. Parker (Producer), *South Park.* New York: Comedy Central.

Buchel, O., & Hill, L. L. (2011). Treatment of georeferencing in knowledge organization systems: North American contributions to integrated georeferencing. *NASKO, 2*(1), 47–57.

Byrne, G., & Goddard, L. (2010). The strongest link: Libraries and linked data. *D-Lib magazine, 16*(11), 5. doi:10.1045/november2010-byrne.

Carroll, L. (1893). *Sylvie and Bruno concluded.* London: Macmillan.

Cutter, C. A., Ford, W. C., Phillips, P. L., & Conneck, O. G. T. (1904). *Rules for a dictionary catalog*. Washington, DC: GPO.

Danko, D. M. (2012). Geospatial metadata. In W. Kresse & D. M. Danko (Eds.), *Springer handbook of geographic information* (pp. 191–244). Berlin: Springer.

Drazniowsky, R. (1966). Bibliographies as tools for map acquisition and map compilation. *Cartographer, 3*, 138–141.

Dublin Core Metadata Initiative. (2015). Retrieved from http://dublincore.org.

Durante, K., & Hardy, D. (2015). Discovery, management, and preservation of geospatial data using hydra. *Journal of Map and Geography Libraries, 11*(2), 123–154. doi:10.1080/15420353.2015.1041630.

Fary, M., & Owen, K. (2013). *Developing an institutional research data management plan service*. Retrievedfromhttp://www.educause.edu/library/resources/developing-institutional-research-data-management-plan-service.

Federal Geographic Data Committee (2000). *Content standard for digital geospatial metadata workbook*. Reston, VA: USGS Retrieved fromhttps://www.fgdc.gov/metadata/documents/workbook_0501_bmk.pdf.

Federal Geographic Data Committee (2004). *NSDI future directions initiative: Towards a national geospatial strategy and implementation plan.* .http://www.fgdc.gov/policyandplanning/future-directions/reports/FD_Final_Report.pdf

Federal Geographic Data Committee. (2015a). Retrieved from http://www.fgdc.gov.

Federal Geographic Data Committee. (2015b). *Geospatial metadata tools*. Retrieved from http://www.fgdc.gov/metadata/geospatial-metadata-tools.

Functional Requirements for Bibliographic Records (1998). *International Federation of Library Associations*. Munich, Germany: K.G. Saur Verlag.

Gilliland, A. J. (2008). Setting the stage. In M. Baca & Getty Research Institute (Eds.), *Introduction to metadata* (pp. 1–19). Getty Research Institute: Los Angeles, CA.

Goodchild, M. (2013). *Personal communication*, April 4, 2013.

Greenberg, J. (2001). A quantitative categorical analysis of metadata elements in image-applicable metadata schemas. *Journal of the American Society for Information Science and Technology, 52*(11), 917–924. doi:10.1002/asi.1170.

Hardy, D., & Durante, K. (2014). A metadata schema for geospatial resource discovery use cases. *Code4lib Journal, 25*. Retrieved from http://journal.code4lib.org/articles/9710.

Hitt, J. (1996, March 10). The theory of supermarkets. *New York Times Magazine*, pp. 56–61.

Kervin, K. E., Michener, W. K., & Cook, R. B. (2013). Common errors in ecological data sharing. *Journal of eScience Librarianship, 2*(2), 3–16. doi:10.7191/jeslib.2013.1024.

Kessler, C., Janowicz, K., & Bishr, M. (2009). *An agenda for the next generation gazetteer: Geographic information contribution and retrieval.* 17th ACM SIGSPATIAL international conference on advances in geographic information systems, Seattle, WA.

Kiszko, K. M., Martinez, O. D., Abrams, C., & Elbel, B. (2014). The influence of calorie labeling on food orders and consumption: A review of the literature. *Journal of Community Health, 39*(6), 1248–1269. doi:10.1007/s10900-014-9876-0.

Klemp, R. P. (1982). On the access to cartographic collections in GDR libraries. *INSPEl, 16*, 31–39.

Kozimor, J., Habermann, T., Gordon, S. Powers, L., & Farley, J. (2016). *Big Earth Data Initiative (BEDI) metadata improvement*: Case studies. Earth Science Information Partners Winter Meeting 2016, Washington, D.C.

Larsgaard, M. L. (1998). *Map Librarianship: An introduction*. Englewood, CO: Libraries Unlimited.

Linnaeus, C. (1758). *Systema Naturae* (10th ed.). Biodiversity Heritage Library OAI Repository: Citebank Retrieved from https://archive.org/details/cbarchive_53979_linnaeus1758systemanaturae1758.

Makri, S. (2014). *Serendipity is not bullshit*. EuroHCIR 2014, The 4th European symposium on human-computer interaction and information retrieval, London, UK.

Martin, J. (1982). *Strategic data planning method*. Upper Saddle River, NJ: Prentice Hall PTR.

Martinolich, K. (2015). *Introduction to geospatial metadata*. Retrieved from ftp://ftp.ncddc.noaa.gov/pub/Metadata/Online_ISO_Training/Intro_to_Geospatial_Metadata/presentations/1-IntroductionToMetadata.pptx.

Medeiros, N., Bills, L., Blatchley, J., Pascale, C., & Weir, B. (2011). Managing administrative metadata. *Library Resources & Technical Services, 47*(1), 28–35. doi:10.5860/lrts.47n1.28.

Michener, W. K., Brunt, J. W., Helly, J. J., Kirchner, T. B., & Stafford, S. G. (1997). Nongeospatial metadata for the ecological sciences. *Ecological Applications, 7*(1), 330–342. doi:10.1890/1051-0761(1997)007[0330:NMFTES]2.0.CO;2.

Monmonier, M. (2008). Web cartography and the dissemination of cartographic information about coastal inundation and sea level rise. In M. P. Peterson (Ed.), *International perspectives on maps and the Internet* (pp. 48–72). Berlin: Springer.

National Information Standards Organization (2004). *Understanding metadata*. Bethesda, MD: NISO Press.

Online Computer Library Center. (2015). Retrieved from http://www.oclc.org/dewey.en.html.

Rothman, R. L., Housam, R., Weiss, H., Davis, D., Gregory, R., Gebretsadik, T., et al. (2006). Patient understanding of food labels: The role of literacy and numeracy. *American Journal of Preventive Medicine, 31*(5), 391–398. doi:10.1016/j.amepre.2006.07.025.

Short, K.C. (2013). *Spatial wildfire occurrence data for the United States, 1992–2011*. [FPA_FOD_20130422]. Fort Collins, CO: USDA Forest Service, Rocky Mountain Research Station. doi:10.2737/RDS-2013-0009.

Sinclair, S. E., Cooper, M., & Mansfield, E. D. (2014). The influence of menu labeling on calories selected or consumed: A systematic review and meta-analysis. *Journal of the Academy Nutrition and Dietetics, 114*(9), 1375–1388. doi:10.1016/j.jand.2014.05.014.

Stephens, S. L. (2005). Forest fire causes and extent on United States Forest Service lands. *International Journal of Wildland Fire, 14*(3), 213–222. doi:10.1071/WF04006.

Szostak, R. (2014). The importance of knowledge organization. *Bulletin of the American Society for Information Science and Technology, 40*(4), 37–42. doi:10.1002/bult.2014.1720400414.

Taylor, A. G., & Joudrey, D. N. (2009). *The organization of information*. Westport, CA: Libraries Unlimited.

Taylor, R. (1968). Question-negotiation and information seeking in libraries. *College and Research Libraries, 29*, 178–194.

United State Geological Survey. (2015). *South Florida Information Access*. Retrieved from http://sofia.usgs.gov/metadata/.

Wood, D. (2010). *Rethinking the power of maps*. New York: The Guilford Press.

Chapter 6
Geoweb

Abstract This chapter reviews the technology, data, and implications related to the Geospatial Web (Geoweb). The unprecedented volume of geographic information (GI) being created, made available, editable, and distributable presents challenges and opportunities for information professionals related to GI access and use. Usability, functionality, and accessibility issues for web-based mapping applications and location-based mobile applications provide one stream of research for Information Science. The Geoweb, as both the cyberinfrastructure and the information space for GI, requires some re-framing of the various resources that support discovery. The contextualization of the user roles, communities, and GI types provide other new avenues of research in Information Science related to human information seeking behavior and the digital curation of all types of information.

6.1 Everywhere, Anyplace

In *Ambient Findability*, Peter Morville (2005) introduced a vision of the future in which "we can find anyone or anything from anywhere at anytime" (p. 6). As an information architect, Morville saw what could be done with the technological building blocks that made a foundation for the Semantic Web (i.e., Web 3.0). The Semantic Web could empower all connected users to share information regardless of their locations in machine-readable formats (e.g., linked data), which allow artificial intelligence to utilize information in transformative ways. Despite many inherent geographical aspects of this ambient findable world (e.g., geospatial ontologies), the Web and its users can be the focus for information architects, web producers, graphic designers, and other individuals working in Human-Computer Interaction (HCI) without any direct consideration of geography (Dix, 2009; Janowicz, Raubal, & Kuhn, 2011). To be clear, the Internet operates as a network of networks with one of its most visible resources being the World Wide Web (Web). The Web allows a common information space to share interlinked hypertext in various formats through standards and protocols overseen in part by the World Wide Web Consortium (W3C). In the U.S. 84% of adults use the Internet (Perrin & Duggan, 2015).

The Geospatial Web (Geoweb) does not exist without the Web, its markup languages, and its associated user footprint. It is also important to consider issues of accessibility (e.g., price, quality of service) in the context of the Web, as well as

W. Bishop, T.H. Grubesic, *Geographic Information*, Springer Geography,
DOI 10.1007/978-3-319-22789-4_6

differently abled users. All of these factors impact the Geoweb, at least in some way. Despite the Greek prefix "geo" coming first when referencing this domain, much of the Geoweb development approaches the "geo" portion as a secondary concern and/or interest, emphasizing the information content first and then using the geographic portion of the data as an interesting display feature. This is unfortunate because there is significant value embedded within the spatial information that is infrequently leveraged when dealing with Geoweb data. Information Science approaches to geographic information (GI) access and GI use *could* (and *should*) take advantage of the opportunity to cross-pollinate with the spatial sciences (broadly defined), especially when it comes to GI organization. The potential research areas listed in this discussion will not represent a comprehensive list of Information Science contributions, nor should future work on GI access and use be confined to the Web alone. The dominance of the Geoweb in the discourse on GI, combined with the sheer volume of GI born digital and available online, give the lens of Information Science a pragmatic platform to build a research framework upon. Information Science work extends beyond the study of information organization to work on user experience design (UXD), information services, human information seeking behavior, and digital curation for all types of data, not just GI.

Today's social software, semi-functioning Sematic webs, and ubiquitous computing, closely match Morville's decade-old prediction. A great deal of information and data pertaining to day-to-day activities and the rhythms of life are widely available from a multitude of sources. Locality may not receive the same prominence in discussions in Information Science as it does in Geography, but the potential geoparsing of most online information, as well as the importance of place for users can no longer be ignored in non-geographic disciplines. There are many potential reasons that GI has been overlooked by other disciplines. Spatial data are difficult to work with, demand a comprehensive understanding of uncertainty and error, need to be abstracted and projected, and so forth. Perhaps a misconstrued sentiment related to the death of distance (Cairncross, 2001) extrapolated to all GI caused many researchers in fields outside of Geography to overlook the geographic aspects of the Web until recently. Whatever the reasons, the historic lack of focus on spatial data is beginning to wain and users now access and create GI about everywhere from anyplace. Locality assumes a crucial role in society even with the growing irrelevance of distance for information transfer (Sui, Goodchild, & Elwood, 2013). The people working in virtual spaces do exist in physical space and its influence on their work (either directly or indirectly) is inescapable. The question of negotiation and information sharing between those geographies are ripe for exploration (Bishop, 2011). A great deal of online information from everywhere also may be disseminated using Geoweb tools in anyplace. This communicational turn places location-based applications (apps) and their maps in the pockets of most everyone. In turn, it is important to consider implications for GI-related usability, functionality, accessibility, discoverability, information needs, information seeking behaviors, and curation throughout the data lifecycle.

The GI access and use discussions about this omnipresent ability to discover all information that stems from this new platform are complex and numerous. As a single chapter introducing an expansive concept with a focus on GI access and use, several branches of the Geoweb tree will be lopped off for intellectual kindling elsewhere. A fuller Geoweb dialogue in another volume should include a review of Internet connectivity speeds and costs (i.e., the digital divide), various literacies required to access and use GI and geo-enabled information (i.e., spatial literacy), and other situational factors that impact how truly *real* Morville's dream has become for all residents of earth (i.e., social justice). In addition, a great deal of geo-enabled information will be dismissed from further discussion for reasons made clear in the next section. For this chapter, we shall presume ambient findability as a default for the GI access and use framework presented. For the benefit of readers new to the Geoweb topic, the following provides a review of the concept, relevant information types, some of the implications resulting from this new epoch of GI access and use, and some of the technological infrastructure of the Geoweb.

6.2 Geoweb

Over the past two decades, devices used to access the Web have become small, affordable, and nearly ubiquitous. The world is shifting away from an Internet of computers toward an Internet of things (IoT) (Mattern & Floerkemeier, 2010). As more devices become connected to the Internet, including thermostats, refrigerators, door locks and lighting systems, the Geoweb will expand rapidly. The anytime and anywhere aspects of the Internet are close to reality. The Geospatial Web, GeoWeb, or Geoweb refers to this intersection between the Web, geospatial technologies, and information (Herring, 1994). However, the Geoweb is not a new concept, nor is it a common referent to this findable, ambient phenomenon outside of academia. Even within Geography, the term Geoweb reflects different aspects of this confluence of information and geospatial technology, as well as big data. For some, the persistence of various interrelated neologisms exists only as an excuse to publish new work. However, it is also important to make efforts to unpack the complexities of the Geoweb, from volunteered geographic information (VGI) (Goodchild, 2007), neogeography (Turner, 2006), and new spatial media (Crampton, 2009). The term Geoweb succinctly captures the broadest concepts of the placial aspects of the Web—media in places and places in media (Adams, 2009).

Sui and Zhao (2015) present a useful dichotomy of GIS evolution for this discussion. It includes "automated cartography (organizing geospatial information) to its current state as media, with an emphasis on the Geoweb (organizing information geospatially)" (p. 191). This dichotomy is most useful as it delineates by the purpose of the information system. Not all Geoweb tools are designed to facilitate automated cartography or the spatial analysis done in traditional GIS packages. The purpose of many Geoweb tools is to organize information—incorporating geography into the presentation of other, online information, and ultimately organizing this information geospatially. A majority of the Geoweb tools that help users' wayfind (e.g., Google Maps), or use social media to check-in, review, and schedule events at locales, or collect personal information that has locality, function without a user having any geographic knowledge. Users navigate and build these metaverses in spatial ways, but the earth's graticule is used as a grid of convenience.

6.2.1 Organizing Information Geospatially

In the information professions, the changes of anytime, anywhere information available through the Geoweb altered the service roles of information providers. The removal of the precondition of being proximally affixed to location-bound technologies to the search for information meant a broader geography was in play. For example, both information seekers and information providers could originate anywhere and the information exchanged became more diverse (in content and location of generation) (McClure & Jaeger, 2009). In most cases, information professionals were not involved in building these new tools and concerns arose over the implications of these new sources given the lack of information literacy skills (e.g., evaluating the authority of sources) for many individuals (Fidel et al., 1999). The Pacific Northwest tree octopus (http://zapatopi.net/treeoctopus/) was an early hoax website used to teach students how to consider trustworthiness and evaluate the authority of online resources. The panic of information without information scientists, especially librarians, parallels similar handwringing of "a geography without geographers" sentiment found in the GIScience literature when the Web went geo (Sui & Zhao, 2008, p. 5). With greater access to these new information retrieval tools that automated searches over vaster amounts of information on the Web, users easily found their answers and met their information needs. These information needs and corresponding location-based questions were not studied because of the assumption that they may be easily answered with minimal training (Katz, 2002).

Location-based questions are inquiries that concern a georeferencable site (Bishop, 2011). To review, georeferencing means "relating information to a geographic location," or in other words, the process of indexing information to places on earth (Hill, 2006, p. 1). Early Geoweb literature highlights the need to geoparse geographic identifiers from existing human artifacts (Scharl, 2006). An early estimate indicated that 20% of Web pages contained recognizable and unambiguous geographic identifiers like physical addresses, telephone numbers, and descriptors of landmarks that allow georeferencing to occur (Delboni, Borges, & Laender, 2005). Many applications modify answers to location-based questions based on location. For example, there is a GIS-based 911 that finds your location based on location-based information on your mobile device.

This Geoweb functionality allows online information to be affixed to a locality and helps answer many everyday life questions. Mobile advertising creates ads based on your location, there are apps to find nearby public restrooms, as well as tools embedded in devices to measure movement velocities and trajectories to help inform local traffic reports. Georeferencing retains a built-in vagueness as witnessed anytime someone provides verbal directions, but through geospatial technologies like global positioning systems (GPS) and geocoding, accurate and precise geographic coordinates can be assigned to many types of information. GPS was initially developed by the U.S. Department of Defense to provide military operations with real-time locational information in the field. It was never intended for civilian use. However, in 1983, President Reagan made a quick policy change that allowed GPS for civilian purposes in response to a Korean airliner being shot down over Soviet Territory (Schneider, 2013). President Clinton removed some purposeful degradation of accuracy and further opened up GPS for civilian and commercial use in 2000. These actions led to a cornerstone of the Geoweb—high accuracy georeferencing that is useful for many basic and/or advanced wayfinding tasks, and other location-based apps (U.S. Department of Defense, 2008).

This ability to navigate urban, suburban, rural, and remote areas, without the assistance of any information providers, was revolutionary. Put simply, for the average user, answers to the majority of all navigational questions no longer required asking anyone for help as long as you had a device and could get a satellite signal. That said, a recent study found that 11.5% of the all questions asked over a three year period at one information agency (129,572) were wayfinding questions and therefore people do still need directions (Bishop, 2012). One of the earliest wayfinding tools on the Geoweb was MapQuest. Launched in 1996, many readers will likely recall printing out driving directions and maps for new destinations. Over time, more vendors began providing services similar to MapQuest and in 2009, Google Maps surpassed it in traffic for location-based needs. By 2013, the Google Maps mobile app became the most used mobile tool on the planet (http://www.complex.com/tech/2013/08/google-maps-most-used-app).

Interestingly, the importance for a human to locate and orient themselves to their environment led to established principles for *You-Are-Here* (YAH) maps that preexist any Geoweb tools (e.g., salient labels in terrain and the map, align map with the terrain, and so forth) (Levine, 1982). With the explosion of a myriad of location-based services through mobile devices, *here* is now where all app users are, and YAH is the default map for most personal uses. The labels in the built environment link to common symbols in location-based service apps' maps. "Location-based services (LBS) are a subset of web services meant to provide functions that are location-aware, where the use of such services is predicated on knowledge of where the services are engaged" (Wilson, 2012, p. 1267). In an effort to meet user needs related to where, over $115 million had been invested in 2009 in 'location start-ups' (Miller & Wortham, 2010). Fulfilling those simple wayfinding needs concerning physical access to place-bound things (e.g., restaurants) restructures everyday life into a series of transactions with digital traces—the result is a conspicuous mobility constructed for the whole of humanity (Wilson, 2012).

Many of these wayfinding aspects of the Geoweb provide new avenues for researchers studying the algorithms that create waypoints and routes. There are other applications for locative social media that can be summarized into three main groups: (1) Social check-in (e.g., the ability to become "mayor" of Casey's in Pittsburgh by checking in more than anyone else on Foursquare); (2) Social review sites (e.g., Yelp helping you avoid bad Thai food in the Midwest); (3) Social scheduling/events sites (e.g., networking with local business and academic people interested in data science through Meetup) (Thielmann, 2010). With many users participating with the social media tools and giving data to produce these digital traces, Foursquare and others have expanded the possibilities for researchers to engage with geo-enabled social informatics. Foursquare reports having 45 million accounts and 5 billion check-ins as of December 2013 (Crowley, 2013). Needless to say, these types of tools and their associated data provide a huge corpus of data to analyze.

Researchers focused on the design, implementation, and use of information and communication technologies (ICTs) for people do not typically consider or understand geographic factors (Sawyer & Rosenbaum, 2000). The geographic aspects of some of the social media tools' digital traces warrant review in relation to human information seeking behavior. However, the usability of social media seems like a moot point. Few users need the assistance of an information professional in finding or using these tools, and even fewer information agencies can support the curation of the access and use to these truly big data. The Library of Congress ingests nearly half a billion tweets per day as of October 2012, but the resources of the largest library in the world cannot be duplicated elsewhere (Library of Congress, 2013). At best, much of the geoparsable social media allows for skewed visualizations of cyberspace for entertainment purposes (Zook et al., 2015). In the new paradigm of data-intensive scientific discovery, many attempt to tease out meaning from big social media data occur (Hey, Tansley, & Tolle, 2009). In 2013, one study of Flickr (which holds nearly 200 million geotagged photographs) compared 836 sites in 31 countries to show how crowd-sourced digital traces may be suitable proxy for the more traditional measures of visitation counts, with the big untested assumption that picture taking and uploading is constant over space (Wood, Guerry, Silver, & Lacayo, 2013). The data and tools make these types of analyses possible, but the research artifacts all inherit the limitations and epistemologies of both.

Additionally, there remains the ocean of personal sound, image, location, and motion data collected through mobile devices that compile physiological factors tracking relationships between stress, mood, food, sex, and sleep of individuals (Hill, 2011). "Participatory personal data are any representation recorded by an individual, about an individual, using a mediating technology" (Shilton, 2012, p. 1906). Aggregating these personal accounts provide another wonderful body of data for health and social science research. However, the geographic factors may be an overvalued aspect of this information. For example, the McGriddle sandwich (http://tinyurl.com/lwmh9w3) is nearly 500 calories, regardless of one's elevation, latitude, or longitude. Personal locality, which is ubiquitous in this geo-enabled information, provides research on human information seeking behavior a means of utilizing geographic factors in their studies (Crawford, Lingel, & Karppi, 2015). The analytical implications of aggregating social data to arbitrary political boundaries are well known (see the modifiable areal unit problem (MAUP) (Openshaw, 1984). But there are also complications associated with uneven data collection of the same variable across varying levels of government and allied agencies. The effects of both are amplified when dealing with personal data. Although these data can be organized geographically, efforts to either elicit or compel coordinated GI access and use efforts from information professionals are rare (Downey, 2006). Interestingly, the devices which collect location-based information *for* and *about* users is fulfilling the users' needs for information about themselves. This is a somewhat difficult concept to wrap one's head around. Consider a related example. With the help of GPS collars, humans have long studied movement patterns of other species (e.g., wolves, bears, and so forth). Today, humans now willingly provide that same information about themselves because it provides locational context about their actual (or desired) activities on the earth. In effect, because the devices are now so good at leveraging this information, humans are freed from retaining spatial memory (Loarie, Van Aarde, & Pimm, 2009). The implications of this may be far-reaching, but the research is on-going.

6.2.2 *Organizing GI*

A great deal of the Geoweb remains after one removes all of the geo-enabled information from location-based services (i.e., wayfinding and social media) and participatory personal information (i.e., activity trackers). Following the Sui and Zhao (2015) dichotomy, the parts of the Geoweb that benefit most from the contributions of information professionals pertain to organizing geospatial information not the organizing of information geospatially. The distinction between GI and geo-enabled information requires one to revisit the core elements of both information and geography. Information can be categorized into three classes: (1) physical and digital information objects (i.e., information-as-thing); (2) communication of information (i.e., information-as-process); and (3) cognition (i.e., information-as-knowledge) (Buckland, 1991). All of these types of information can be geographic. Throughout this text, however, we have primarily explored GI using the information-as-thing definition. For example, all cartographic resources that are organized into collections and databases are measurable objects that can receive descriptive metadata. These same cartographic resources, on the Geoweb (or not), can also operate under the information-as-process definition, especially if one explores the map communication model (MCM) popularized by Arthur Robinson during the scientification of cartography (Crampton, 2001). Readers may find it intriguing that the MCM evolved from the same Shannon-Weaver model of communication that informed early efforts in Information Science (Shannon & Weaver, 1949). In a nutshell, a cartographic resource may be learned, and the cognition that occurs to encrypt that geospatial data into the brain is GI-as-knowledge.

As detailed throughout this book, GI can take the form of objects, or GI-as-thing. For example, the position on the earth's surface can be the attribute of interest. However, when considering the relative importance of location in the broader context of geo-enabled information, it may be of secondary interest, or in some cases inconsequential. For instance, a Flickr hosted photograph of Cinderella's castle at the Magic Kingdom Theme Park in Lake Buena Vista, Florida includes latitude and longitude coordinates (28.4195 °N, 81.5812 °W), but this GI is likely not relevant to the information needs or information seeking behaviors for that information object.

Conversely, wayfinding does rely entirely on GI to function, but the raw, GPS-derived information is converted and/or transposed into turn-by-turn directions related to the salient labels and terrain of surroundings. This process is done with all of the relevant geographic aspects present, but they are not absolutely crucial for success. For example, the spatial cue may be an instruction to "turn right on Siskiyou Boulevard". The geographic cue will include information as to where that turn needs to occur (e.g., 500 ft). In this instance, both pieces of information are useful, but successful navigation could occur with only the spatial cue.

Of course, the Geoweb supports the access and use of many GI resources where the geographic elements meet specific information needs. For example, elements of precision, accuracy, coordinate systems, and projections all matter in determining fitness for use. To clarify, these facets of GI exclude tools that address relationships between origins (i.e., here) to destinations (i.e., there), visit frequencies to the origins and destinations, tabulations of the specific people or objects that have visited a location, or information concerning any type of physiological interactions in a specific location. These remaining GI resources all correspond to more complex location-based questions concerning attributes of geography. All of them live within the domain of GI.

One of the more significant benefits accrued by Geoweb resources is that GI is born from many sources. For example, in 2013, Open Street Map reached one million registered users. The information injected into these types of tools represents information that transcends everyday life activities and can be used to help decision-making, as well as the creation of place (Zook & Graham, 2007). Unlike many of the wayfinding information needs of social media, there is no killer-app that provides a quick, automated response to the types of location-based questions related to these GI. It is far too complex and nuanced. Therefore, a significant number of opportunities exist for information professionals to serve as intermediaries for these types of queries. They can shepherd users to GI resources that meet information needs and promote use. For example, several tools exist for the mapping of historically underrepresented, neglected communities, and their associated geographies. For example, Harrassmap (http://harassmap.org/en/) (Leszczynski & Elwood, 2015) details occurrences of sexual harassment and assault. Information professionals could build guides to list these types of tools or refer users to them when asked a location-based question related to that type of GI. Information that has been georeferenced or even geocoded, but done so without the traditional concerns of geographic representation, would impact future use. But, geographic aspects still qualify those GI as information where geography matters the most.

The Geoweb allows local knowledge defined as "practical, collective and strongly rooted in a particular place" and that forms an "organized body of thought based on the immediacy of experience" to be shared (Geertz, 1983, p. 75). Although Geertz was studying cultural norms across places, the concept of local knowledge is relevant to the Geoweb in the context of information providers and users. Local knowledge, also known as indigenous knowledge, may now be shared with all information providers and users to inform GI about a place. Knowledge related to technologies, skills, practices, and beliefs, which enable a community to establish a stable existence in their environment, comprise local knowledge (United Nations, 2008). The United Nations utilizes indigenous people's local knowledge to assist in nature conservation, disaster management, and to preserve traditional medical practices. This type of GI is critical to providing answers to all types of location-based questions beyond more traditional GI sources (e.g., U.S. Geological Survey (USGS) topographical maps). Large portions of this GI have been produced through VGI channels, which represents a major overall shift in production tactics and methods. Rather than being produced by the few (e.g., government sources), GI is now produced by the masses and represents user-generated content mediated by mobile devices. The increasing accessibility and popularity of VGI have had ramifications in conventional notions of mapmaking and geography, including the ways that configurable online maps invite questions of political subjectivity and legitimate versus illegitimate participation (Parks, 2009). Indeed, VGI is increasingly being used to supplement knowledge gaps in emergency response and disaster management (Graham & Zook, 2011), citizen science (Goodchild, 2007), and even helping build the National Map (Poore, Wolf, Korris, Walter, & Matthews, 2012). Beyond crisis informatics, academics, and other information professionals now recognize that the many ways in which user–generated content on Geoweb platforms could be utilized to help answer long–standing research questions requiring empirical evidence at a grand scale (Haklay, Singleton, & Parker, 2008). To facilitate all these processes, Information Science could assist with the usability, functionality, accessibility, discoverability, information needs, information seeking behaviors, and curation through the data lifecycle.

One of the complicating factors associated with the Geoweb is that the line between authoritative, professional, and scientific analysis and visualization gets somewhat muddled. The sheer variety of data accessible for use is intoxicating, but this array of options does not always lead to a clear choice for authority control. The populist maps produced by the crowd now appear both scientific and authoritative, even when they lack the underlying data quality and analytical rigor necessary to qualify as such. Rather, these projects might be considered more democratic, more amateur, and non-rigorous compared to something created by the USGS, but if ground truth reflects local knowledge, then these types of GI *add value and fill in gaps* not covered by government and private sector resources. Chapter 7 will outline a few of the more nuanced GI access and use issues related to discovering authoritative GI (i.e., geospatial data, digital maps, and so forth). To reiterate, Geoweb GI is sometimes less formal, less authoritative, volunteered, unwittingly given, or culled from big data sources. However, this GI benefits from the work of information professionals to facilitate access and use through the study of its usability, available GI resources, and the associated human information seeking behaviors, as well as issues inherent to access and use themselves (Bishop, Grubesic, & Prasertong, 2013).

6.3 Implications

The rise of the Geoweb presents new challenges for information professionals as information intermediaries to the plurality of geographic representations. Huge volumes of GI is a good problem to have, but the increased volume means that modifications to information policy, digital curation, equitable access, and usability must be considered. Much of Chap. 4 (Policy) discussed U.S. government information, but the policies related to costs, privacy, and security expand to all Geoweb GI. Policy instruments in the United States, the United Kingdom, Australia and all other countries impact GI access and use. For example, policy for GI organization impacts what GI is created and collected by various government agencies, but also what GI is created and collected by every citizen. For all GI, the most salient policy implications raised relate to individual privacy and security. The same Geoweb tools that make wayfinding easier, social media engaging, and participatory personal data useful, provide an unfiltered window into one's personal life, easily mined by service providers, the National Security Agency or other organizations that may not respect or value individual privacy the way that users might appreciate. Information professionals can both remain vigilant in following policy changes and educate users of the impacts to their GI access and use. In comparison with information policy implications, information professionals play a much larger role in the consequences of digital curation as stewards of the human record.

Chapter 9, which explores the data lifecycle and details issues related to the digital curation, will focus on traditional GI. Within the Geoweb, data quality and accuracy, accessibility, authorship, and authority all start to blur. As a result, this often makes the metadata creation tasks of information professionals as incomplete as the data itself. Without detail upon creation, subsequent GI, and geo-enabled information lacks the data provenance necessary to ensure future use and inform reuse. Provenance and uncertainty of different sources should be maintained in synthesis, but often is not in these emergent fields of data science and GIScience big data communities. Emerging issues for the Geoweb include the implications of aggregating GI and geo-enabled information of various accuracies, at different levels of detail, and from different generalizations. These are very good questions for informatics experts, as well as GIScientists as they are more qualified to address them than any other information professional (Sui & Goodchild, 2011). Yet, the long-term preservation issues and reuse aspects of the data lifecycle do fall into the Information Science wheelhouse.

6.3.1 Accessibility

In their infancy, early Geoweb tools were collectively referred to as "web-based mapping applications". Again, this was a major breakthrough because the Web allowed for greater access to GI and the use of the Internet was more cost (and time) effective when compared to shipping hard drives between places. In fact, the emergence of web-based mapping applications (and later, location-based apps) allowed users without any formal training or access to desktop GIS tools to use GI. The lack of cartographic and geographic knowledge, as well as GIS training for the average end-user, created new challenges for equitable access (Tsou & Curran, 2008).

As detailed earlier, MapQuest, Google Maps, and Google Earth are all prime examples of how access to GI has become more equitable, pervasive, and user-friendly (Roche *et al.*, 2011). Many other Geoweb tools open their platforms (including the underlying code) to allow for an injection of VGI from their users. This allows for a suite of diverse questions to be asked and compelling research topics to be tackled, including elements of political participation, labor ethics, privacy concerns, and archival logistics. Throughout this entire process, information professionals *can* and *do* facilitate access and use of important GI, through organization, and make many contributions behind the scenes (Elwood, Goodchild, & Sui, 2013). Interestingly, most non-governmental Geoweb tools rely on the internal scaffolding of governments' investment in creating GI—the same super-structure that serves basemaps and supports most of the Geoweb. For example, any data found in Google Maps builds off the original data from the U.S. Census Bureau Topologically Integrated Geographic Encoding and Referencing (TIGER) files. The satellite imagery of Google Maps is from the USGS (Madrigal, 2012). Apps beyond location-based services, many of which are common in Google Maps, are possible with more access to open government data. The amount of data from open and freely available government GI is vast due to value added by a myriad of GI vendors, but equitable access presents all users the ability to share, edit, and reimagine their geographies via the Geoweb.

Perhaps the most common example of equitable access and its associated implications is the map mashup. On the face of it, the map mashup appears as a neat parlor trick, enabling any georeferenceable web-based information to be put on a map. Behind the scenes, however, a great deal of geocyberinfrastructure is necessary to support such efforts. We can all thank Paul Rademacher, who created the first map mashup in 2005 (http://housingmaps.com/). To solve his personal residential search needs, Paul did what any programmer would do in his or her free time and built a mashup using Google Maps' open application program interface (API) and combined it with housing data. After this initial effort, a proliferation of map mashups emerged. Today, Google Earth has the same type of functionality and the Keyhole Markup Language (KML), which makes mashups possible, have exploded in popularity. This type of equitable access and creation are not confined to the private sector. For example, *The National Map Corps* (*TNM*Corps) (http://nationalmap.gov/) empowers citizens to collect structures data by adding features, removing obsolete points, and correcting existing data the in The National Map database (USGS, 2016). Also, some information agencies have been able to digitize map collections and mashup with modern Geoweb tools for unprecedented access such as the New York Public Library Map Warper (http://maps.nypl.org/warper/) (Knutzen, 2013).

In the end, equitable access to GI, which results from the Geoweb, is not confined to mashups and VGI. With online transportation network companies such as Uber using the Geoweb to inform routes, passengers are able to see the suggested route from the app and both the drivers and passengers trust the technology to provide good (and equitable) information. In the past, these types of resources were not available to passengers that rode with traditional taxi services/drivers, where route choice was dependent on the local knowledge of the driver. Deviations from the optimal route, whatever that may have been, may or may not have saved time. Moreover, longer routes benefit the drivers by costing the passengers more (Noulas, Salnikov, Lambiotte, & Mascolo, 2015). There are plenty of unpacked authority issues and motivations behind the creation and use of all GI, but at least with the Geoweb, equity in access is growing. More importantly, there is room for improvement and growth and access and equity—more usability studies would be helpful, and additional options for re-design should be explored.

6.3.2 Usability

One of the major conundrums associated with the Geoweb is that geographic philosophy and thought is not required by the designers to study the influence of the Web on human work and activities, spatial or aspatial. Rightly or wrongly, all methods of UXD rely upon input from users to maximize effectiveness, efficiency, and satisfaction of a tool in context. With user expectations and Web technologies both in flux, HCI operates with non-context dependent guidelines, principles, or heuristics rather than evidence-based laws found in some other disciplines. Here is one example of interface design rules:

- Strive for consistency
- Enable frequent users to use shortcuts
- Offer informative feedback
- Design dialog to yield closure
- Offer simple error handling
- Permit easy reversal of actions
- Support internal locus of control
- Reduce short-term memory load (Shneiderman, Plaisant, Cohen, & Jacobs, 2014).

These pragmatic design considerations rely on iterative user feedback to inform interface design. In all UXD, the participating individuals' assumptions and distortions of geographic space, as well as their core understanding of GI, impact their performance and reactions to usability. Participants in usability testing of any GIS should certainly be familiar with GI, its uses, and expected functions (e.g., zooming to scale). Again, basic spatial literacies remain irrelevant for the actual use of many websites, mobile apps, data portals, and numerous other digital resources. But, in the design of many GI tools, the potential audience may not have any geographic and/or cartographic competencies. As a result, the safest level of expertise designers should assume for their intended users would be in the realm of geographic naiveté.

More formally, Naïve Geography is a field of study that attempts to model how people think about geography and incorporate common sense design into GIS, but the same principles could be used for other GI-related tools (Egenhofer, & Mark, 1995). The lack of GIS training, geographic and cartographic knowledge, and the variety of users' geographic information literacies makes creating any user-friendly design difficult for either GIS or other tools serving geo-enabled information. Sadly, for much of the Geoweb, the impetus to improve usability is more users and more revenue from these users. Much of this is also tied to the amount and quality of personal data and information shared. That said, there are a number of compelling democratic reasons (and related policy) that force government GI to be as open as possible.

Not surprisingly, government websites were quick to take advantage of the Geoweb. By 1997, around 88% of local governments had adopted some form of GIS to disseminate GI data to residents (Ganapati 2011, p. 426). In 2005, it was estimated over 60% of municipal governments included some form of interactive GIS on municipal websites for users to view GI (Kaylor, 2005, p. 21). These web-based GIS systems were accessible to a wide range of people but required local public agencies to spend significant monies on implementing and maintaining these systems and data (Kingston, 2007). In this context, the usability issues are similar to those outlined previously. Today, although government GI is more accessible than ever, its overall usability is somewhat difficult to determine (Cöltekin, Heil, Garlandini, & Fabrikant, 2009). The majority of e-government literature largely focuses on results and performance testing, with much less effort dedicated to determining the usability of software portals and the creation of meaningful measurement metrics (Baker, 2009; Tsai, Choi, & Perry, 2009). Moreover, most Geoweb tools still lack common standards for graphical user interfaces, syntax, or vocabulary, but progress is occurring with new implementation standards (You, Chen, Liu, & Lin, 2007; http://www.opengeospatial.org/docs/is). Similarly, there are efforts underway to better determine user expectations for interacting with government GI (Bishop, Haggerty, & Richardson, 2015).

The good news is initial struggles for government applications can be entirely avoided with today's Geoweb technology with additional design considerations informed by usability testing. In addition, a one size fits all tool for the entirety of GI, or even government GI may never be practical as most users have limitations of understanding the geographic categorization of the world around them. The following reviews the technological marvel that enables us to share everything from UFO sightings to the spread of white-nose syndrome in bats.

6.4 Geocyberinfrastructure

One last component of the Geoweb worth reviewing is the geocyberinfrastructure. Although it is relatively clear that a combination of technological advancements, including the Web have enabled researchers to access any and all information more quickly and readily, a review of the actual technologies supporting this increased level of accessibility is necessary. In other words, it is important to document the "back end" of the Geoweb, especially for those information professionals interested in designing tools for this domain. A good place to start is by examining the importance of peer-to-peer (ptp) networks. Ptp networks allow for a group of computers to form a self-organized network and users to share any information more easily and more quickly than that of earlier types of data transfer (e.g., CDROM) (Fox, Suryanata, Hershock, & Pramono, 2005). As more users access GI and geo-enabled information through the Web, the static nature of early Web pages and relational databases were forced to adapt, given the constant change expected in Web 2.0. To make small changes to a Web page or geospatial dataset required developers to upload and overwrite an entirely new version of the page or dataset instead of simply editing the small changes into a dynamic document (Stamey, Lassez, Boorn, & Rossi, 2007). The temporal lag between full updates caused frustration for developers and created versioning issues that led to a multitude of problems in the curation of GI (Peng, 2005). In addition to these versioning issues and the temporal lag, users could only share information that was interoperable with other users' systems. These two needs of real-time updates and overall interoperability required advances, standards, and compliance, to make Web 2.0, and the Geoweb work.

AJAX (Asynchronous JavaScript) along with EXtensible Markup Language (XML), and other scripts and codes, standards, and programs were soon implemented. These tools allowed for small changes to components of Web pages and associated data/GI without the time-consuming process of uploading an entirely new page or dataset (Mesbah & van Derusen, 2007). As a result, Web traffic increased and the new capabilities became commonly referred to as Web 2.0 (O'Reilly, 2009). The dynamic interfaces and increased traffic generated two major concerns for real-time information delivery: (1) network latency (i.e. the measure of time between sending and receiving a message transferred from a server), and (2) the real-time accuracy of all geospatial data being uploaded and downloaded.

Prior to the introduction of these Web 2.0 technologies, GI often had proprietary image formats and database schemes (Carvalho, Masiero, Lemos, Vera, & Metello, 2007). Sharing GI was complicated due to a lack of interoperability between different organizations and/or between individuals within the same organization. In response to this earlier issue, the Open Geospatial Consortium (OGC) (http://www.opengeospatial.org/) formed in 1994 and began to build the standards that would enable more streamlined sharing of geospatial data through the Web (e.g., Geography Markup Language (GML), Web Feature Service (WFS), and Scalable Vector Graphics (SVG) (Peng & Tsou, 2003).

GML is an XML language for encoding GI's metadata. GML allows users to define feature-based relationships between different geographic databases, regardless of the network arrangement or GI format (Gerlek & Fleagle, 2007). GML code is structure independent, which increases its interoperability with various GIS packages (Lake, Burggraf, Trninic, & Rae, 2004). GML also allows users to code at the feature level (i.e., road name or physical attribute). Thus, when changes are made in one geodatabase, it can result in a simultaneous change in all other geodatabases relying on that database for feature information. When combined with APIs and many of the more emergent geoplatforms, this fusion of technology fueled the rapid growth of the Geoweb. In addition to the dynamic ability to update datasets without the time-consuming process of replacing entire datasets, GML, through its XLink function, also allows users to describe relationships between two or more features in a dataset. For example, if construction on one road segment causes a bus route to change, the change would be automatically reflected in all Web sites dedicated to displaying the bus routes and all geodatabases that utilize the bus route in traffic analysis. Building this type of intelligence into information systems is tedious and complicated, but the relative successes associated with doing this highlight the vast potential of GML in realizing the benefits of real-time information delivery for many digital domains.

The Web Feature Service (WFS) is another component worth detailing. WFS allows GML to be queried, which enables remote access to and the manipulation of geodatabases. This interoperable platform allows researchers to synchronize geodatabases, produce map visualizations of GI, and support end-user applications (Lake & Farley, 2007). WFS also allows databases to subscribe or link to other geodatabases, regardless of the underlying database platform. Although WFS technology allows users to synchronize their data more easily (i.e., one change in a database can be immediately reflected in all interlinked databases), the implementation of this format remains a challenge for information professionals. The major benefit of WFS is its removal of interoperability barriers related to real-time information delivery and management. Another OGC standard to facilitate real-time delivery of GI is Scalable Vector Graphics (SVG). Due to the text-based nature of GML, OGC required the creation of a standard format that would allow for the display of GI (Chang & Park, 2006). Through SVG, geospatial data can appear on web browsers and users can query and manipulate that GI remotely. In part, this answers the call for enabling different organizations to view spatial data across systems.

When taken together, the combined use of GML, WFS, SVG, and other standards from OGC alleviated early challenges to real-time information delivery on the Geoweb. The vision of synchronous interoperability between all geodatabases is possible through this geocyberinfrastructure, but challenges to full implementation remain. Specifically, a fuller framework for the implementation of the Geoweb would require more attention and education of these technologies, the adoption of standardized metadata semantics from the Federal Geographic Data Committee (FGDC) (http://www.fgdc.gov/) to support the Semantic Web, some removal of institutional barriers to sharing GI, and the development of more findable and usable tools for non-expert end-users. Clearly, this is a tall order, but technological advancements are rapid. As one can imagine, however, advancements in technology do not alleviate institutional barriers (e.g. lack of interpersonal communication or information hoarding) (Obermeyer & Pinto, 2008). The policy and law are always a few steps behind technology and its users. In some instances, the goals of different organizations may conflict and discourage GI sharing. Changing organizational cultures to view data-sharing delivery as routine and advantageous for all depends on the nature of each organization and the embedded information professionals' history as stewards of the human record. The future of the Geoweb will depend on these characteristics.

6.5 Conclusion

The chapter provided a review of the Geoweb concept, including the geo-enabled information and GI available via the Web. The unprecedented volume of GI and other geo-enabled information being created, made available, and editable presents several implications for information professionals related to GI access and use. Usability, functionality, and accessibility issues for web-based mapping apps and location-based mobile apps provide one stream of research for Information Science outlined in this chapter. The contextualization of the user roles, communities, and GI types that provide other new avenues of research in Information Science, especially as it relates to human information seeking behavior and the curation of data, is especially important for GI and the Geoweb. Information Science work with GI easily extends beyond research concerning information policies and the metadata behind GI organization. In fact, it includes most of the areas presented throughout Part II of this book.

Next, Chap. 7 reviews traditionally authoritative GI resources. Although the discovery issues presented in Chap. 7 will impact all GI in the Geoweb, professional and authoritative geographic representations require a more thorough review to determine fitness for use. Chapter 8 provides a background on human information seeking behavior that begins to delineate roles, communities, and types of use of GI to inform a systematic study of those phenomena. Chapter 9 provides a detailed presentation of the data lifecycle models with examples.

References

Adams, P. C. (2009). Geographies of media and communication. Chichester: Wiley-Blackwell.

Baker, D. L. (2009). Advancing e-government performance in the United States through enhanced usability benchmarks. *Government Information Quarterly, 26*(1), 82–88.

Bishop, B. W. (2011). Location-based questions and local knowledge. *Journal of the American Society for Information Science and Technology, 62*(8), 1594–1603. doi:10.1002/asi.21561.

Bishop, B. W. (2012). Analysis of location-based questions to inform mobile library applications. *Library and Information Science Research, 34*(4), 265–270. doi:10.1016/j.lisr.2012.06.001.

Bishop, B. W., Grubesic, T. H., & Prasertong, S. (2013). Digital curation and the GeoWeb: An emerging role for geographic information librarians. *Journal of Map & Geography Libraries, 9*(3), 296–312. doi:10.1080/15420353.2013.817367.

Bishop, B. W., Haggerty, K. C., & Richardson, B. E. (2015). Usability of E-government mapping applications: Lessons learned from the US National Atlas. *International Journal of Cartography, 1*(2), 1–17. doi:10.1080/23729333.2015.1093333.

Buckland, M. (1991). Information as thing. *Journal of the American Social of Information Science, 42*(5), 351–360.

Cairncross, F. (2001). *The death of distance: How the communications revolution is changing our lives*. Boston: Harvard Business Press.

Carvalho, M. T. M., Masiero, L. P., Lemos, L. P., Vera, M. S., & Metello, M. G. (2007). Continuous interaction with TDK: Improving the user experience in Terralib. *Brazilian Symposium on GeoInformatics*. (pp. 13–22).

Chang, Y. S., & Park, H. D. (2006). XML Web Service-based development model for Internet GIS applications. *International Journal of Geographical Information Science, 20*(4), 371–399.

Cöltekin, A., Heil, B., Garlandini, S., & Fabrikant, S. I. (2009). Evaluation the effectiveness of interactive map interface designs: A case study integrating usability metrics with eye-movement analysis. *Cartography and Geographic Information Sciences, 36*(1), 5–7.

Crampton, J. (2001). Maps as social constructions: Power, communication and visualization. *Progress in Human Geography, 25*(2), 235–252.

Crampton, J. W. (2009). Cartography: Maps 2.0. *Progress in Human Geography, 33*(1), 91–100.

Crawford, K., Lingel, J., & Karppi, T. (2015). Our metrics, ourselves: A hundred years of self-tracking from the weight scale to the wrist wearable device. *European Journal of Cultural Studies, 18*(4–5), 479–496.

Crowley, D. (2013). *Ending the year on a great note (And with a huge thanks and happy holidays to out 45,000,00-strong community)*. Retrieved from http://blog.foursquare.com/post/70494343901/ending-the-year-on-a-great-note-and-with-a-huge

Delboni, T. M., Borges, K. A. V., Laender, A. H. F. (2005). Geographic web search based on positioning expressions. *Proceedings of the 2005 workshop on geographic information retrieval*. Bremen, Germany.

Dix, A. (2009). Human–Computer interaction. In L. Liu & M. T. ÖZsu (Eds.), *Encyclopedia of database systems* (pp. 1327–1331). Boston: Springer.

Downey, L. (2006). Using geographic information systems to reconceptualize spatial relationships and ecological context. *American Journal of Sociology, 112*(2), 567–612.

Egenhofer, M. J., & Mark, D. M. (1995). Naive Geography. In A. U. Frank & W. Kuhn (Eds.), *Spatial information theory: A theoretical basis for GIS* (pp. 1–15). Berlin: Springer.

Elwood, S., Goodchild, M. F., & Sui, D. (2013). Prospects for VGI research and the emerging fourth paradigm. In D. Sui, S. Elwood, & M. F. Goodchild (Eds.), *Crowdsourcing geographic knowledge: Volunteered geographic information (VGI) in theory and practice* (pp. 361–375). New York: Springer. doi:10.1300/J104v41n01_05.

Federal Geographic Data Committee. (2016). *Home*. Retrieved from http://www.fgdc.gov/

Fidel, R., Davies, R. K., Douglass, M. H., Holder, J. K., Hopkins, C. J., Kushner, E. J., Miyagishima, B. K., & Toney, C. D. (1999). A visit to the information mall: Web searching behavior of high school students. *Journal of the American Society for Information Science, 50*(1), 24–37.

Fox, J., Suryanata, K., Hershock, P., & Pramono, A. H. (2005). Mapping power: Ironic effects of spatial information technology. In J. Fox, K. Suryanta, & P. Hershock (Eds.), *Mapping communities: Ethics, values, practice* (pp. 1–11). Honolulu, Hawaii: East-West Centre.

Ganapati, S. (2011). Uses of public participation geographic information systems applications in e-government. *Public Administration Review, 71*(3), 425–434. doi:10.1111/j.1540-6210.2011.02226.x.

Geertz, C. (1983). *Local Knowledge*. New York: Basic Books.

Gerlek, M. P., & Fleagle, M. (2007). Imaging on the geospatial web using JPEG 2000. In A. Scharl & K. Tochtermann (Eds.), *The geospatial web—How geobrowers, social software and web 2.0 are shaping the network society* (pp. 27–38). London: Springer.

Goodchild, M. F. (2007). Citizens as sensors: The world of volunteered geography. *GeoJournal, 69*(4), 211–221. doi:10.1007/s10708-007-9111-y.

Graham, M., & Zook, M. A. (2011). Visualizing global cyberscapes: Mapping user-generated placemarks. *Journal of Urban Technology, 18*(1), 115–132.

Haklay, M., Singleton, A., & Parker, C. (2008). Web mapping 2.0: The neogeography of the GeoWeb. *Geography Compass, 2*(6), 2011–2039.

Herring, C. (1994). *An architecture of cyberspace: Spatialization of the Internet*. Champaign, IL: Army Construction Engineering Research Laboratory.

Hey, A. J. G., Tansley, S., & Tolle, K. M. (2009). *The fourth paradigm: Data-intensive scientific discovery*. Redmond, WA: Microsoft Research.

Hill, K. (2011). Adventures in self-surveillance: Fitbit, tracking my movement and sleep. *Forbes*. Retrieved from http://blogs.forbes.com/kashmirhill/2011/02/25/adventures-in-self-surveillance-fitbit-tracking-my-movement-and-sleep/

Hill, L. L. (2006). *Georeferencing: The geographic associations of information*. Boston: The MIT Press.

Janowicz, K., Raubal, M., & Kuhn, W. (2011). The semantics of similarity in geographic information retrieval. *Journal of Spatial Information Science, 2*, 29–57.

Katz, W. (2002). *Introduction to reference work* (8th ed.). Boston: McGraw-Hill.

Kaylor, C. H. (2005). The next wave of e-government: The challenges of data architecture. *Bulletin of the American Sociaty for Information Science and Technology, 31*(2), 10–22.

Kingston, R. (2007). Public participantion in local policy decision-making: The role of web-based mapping. *Cartographic Journal, 44*(2), 138–144.

Knutzen, M. A. (2013). Unbinding the atlas: Moving the NYPL map collection beyond digitization. *Journal of Map & Geography Libraries, 9*(1–2), 8–24.

Lake, R. & Farley, J. (2007). Infrastructure for the geospatial web. In *The geospatial web—How geobrowers, social software and web 2.0 are shaping the network society* (pp. 15–26). London: Springer.

Lake, R., Burggraf, D., Trninic, M., & Rae, L. (2004). *Geography mark-up language: Foundation for the Geo-Web*. New York: Wiley.

Leszczynski, A., & Elwood, S. (2015). Feminist geographies of new spatial media. *The Canadian Geographer/Le Géographe canadien, 59*(1), 12–28. doi:10.1111/cag.12093.

Levine, M. (1982). You-are-here maps: Psychological considerations. *Environment and Behavior, 14*(2), 221–237. doi:10.1177/0013916584142006.

Library of Congress (2013). *Update on the Twitter archive at the Library of Congress*. Washington, DC: Library of Congress.

Loarie, S. R., Van Aarde, R. J., & Pimm, S. L. (2009). Fences and artificial water affect African savannah elephant movement patterns. *Biological Conservation, 142*(12), 3086–3098.

Madrigal, A. C. (2012). How Google builds its maps—And what it means for the future of everything. *The Atlantic*. Retrieved from http://www.theatlantic.com/technology/archive/2012/09/how-google-builds-its-maps-and-what-it-means-for-the-future-of-everything/261913/

Mattern, F., & Floerkemeier, C. (2010). From the internet of computers to the internet of things. In S. Kai, P. Ilia, & G. Pablo (Eds.), *From active data management to event-based systems and more* (pp. 242–259). Berlin: Springer.

Mesbah, A. & van Deursen, A. (2007). An architectural style for Ajax. In *WICSA '07: Proceedings fo the 6th Working IEEE/IFIP Conference on Software Architecture* (pp. 44–53). IEEE Computer Society.

McClure, C. R., & Jaeger, P. T. (2009). *Public libraries and Internet service roles: Measuring and maximizing Internet services*. Chicago: American Library Association.

Miller, C. C., & Wortham, J. (2010, August 30). Technology aside, most people still decline to be located. *New York Times*. Retrieved from http://www.nytimes.com/2010/08/30/technology/30location.html?_r=0

Morville, P. (2005). *Ambient Findability*. Sebastopol, CA: O'Reilly Media.

Noulas, A., Salnikov, V., Lambiotte, R., & Mascolo, C. (2015). Mining open datasets for transparency in taxi transport in metropolitan environments. *EPJ Data Science, 4*(1), 1–19.

Obermeyer, N. J., & Pinto, J. K. (2008). *Managing geographic information systems*. New York: The Guilford Press.

Open Geospatial Council. (2016). *OGC Standards*. Retrieved from http://www.opengeospatial.org/docs/is

Openshaw, S. (1984). The modifiable areal unit problem. *Concepts and Techniques in Modern Geography, 38*, 41.

O'Reilly, T. (2009). *What is web 2.0?: Design patterns and business models for the next generation of software*. Sebastopol, CA: O'Reilly Media.

Parks, L. (2009). Digging into Google Earth: An analysis of "Crisis in Darfur". *Geoforum, 40*(4), 535–545. doi:10.1016/j.geoforum.2009.04.004.

Peng, Z. R. (2005). A proposed framework for feature-level geospatial data sharing: A case study for transportation network data. *International Journal of Geographic Information Science, 19*(4), 459–481.

Peng, Z. R., & Tsou, M. H. (2003). *Internet GIS: Distributed geographic information services for the Internet and wireless networks*. Hoboken, NJ: Wiley.

Perrin, A., & Duggan, M. (2015). Americans' internet access: 2000–2015. *Pew Research Center, 26*.

Poore, B. S., Wolf, E. B., Korris, E. M., Walter, J. L., & Matthews, G. D. (2012). *Structures data collection for The National Map using volunteered geographic information*. Reston, VA: USGS Retrieved from http://pubs.usgs.gov/of/2012/1209.

Roche, S. L., Sherman, D. L., Dissanayake, K., Soucy, G., Desmazieres, A., Lamont, D. J., Peles, E., Julien, J. P., Wishart, T. M., Ribchester, R. R., Brophy, P. J., & Gillingwater, T. H. (2011). Loss of Glial Neurofascin155 delays developmental synapse elimination at the neuromuscular junction. *The Journal of Neuroscience, 34*(38), 12904–12918.

Sawyer, S., & Rosenbaum, H. (2000). Social informatics in the information sciences: Current activities and emerging direction. *Informing Science, 3*(2), 89–95.

Schárl, A. (2006). Catalyzing environmental communication through evolving internet technology. *The Environment Communication Yearbook, 3*, 235–242.

Schneider, D. (2013). You are here. *IEEE Spectrum, 50*(12), 34–39. doi:10.1109/MSPEC.2013.6676994.

Shannon, C. E., & Weaver, W. (1949). *A mathematical model of communication*. Urbana, IL: University of Illinois Press.

Shilton, K. (2012). Participatory personal data: An emerging research challenge for the information sciences. *Journal of the American Society for Information Science and Technology, 63*(10), 1905–1915. doi:10.1002/asi.22655.

Shneiderman, B., Plaisant, C., Cohen, M., & Jacobs, S. M. (2014). *Designing the user interface: Strategies for effective human-computer interaction*. Harlow, England: Pearson Education Limited.

Stamey, J., Lassez, J., Boorn, D., & Rossi, R. (2007). Client-side dynamic metadata in web 2.0. *Proceedings of the 25th annual ACM international conference on Design of communication*. (pp. 155–161).

Sui, D., & Goodchild, M. (2011). The convergence of GIS and social media challenges for GIScience. *International Journal of Geographical Information Science, 25*(11), 1737–1748.

Sui, D., & Zhao, B. (2008). The wikification of GIS and its consequences: Or Angelina Jolie's new tattoo and the future of GIS. *Computers, Environment and Urban Systems, 32*(1), 1–5.

Sui, D., & Zhao, B. (2015). GIS as media through the Geoweb. In P. S. Mains, J. Cupples, & C. Lukinbeal (Eds.), *Mediated geographies and geographies of media* (pp. 191–208). Dordrecht, The Netherlands: Springer.

Sui, D. Z., Goodchild, M. F., & Elwood, S. (2013). Introduction: Volunteered geographic information, the exaflood, and the growing digital divide. In D. Z. Sui, S. Elwood, & M. F. Goodchild (Eds.), *Crowdsourcing geographic knowledge: Volunteered geographic information (VGI) in theory and practice* (pp. 1–14). New York: Springer.

Thielmann, T. (2010). Locative media and mediated localities: An introduction to media geography. *Aether: The Journal of Media Geography, 5*(1), 1–17.

Tsai, N., Choi, B., & Perry, M. (2009). Improving the process of e-government initiative: An in-depth case study of web-based GIS implementation. *Government Information Quarterly, 26*(2), 368–376.

Tsou, M., & Curran, J. M. (2008). User-centered design approaches for web mapping applications: A case study with USGS hydrological data in the United States. In M. P. Peterson (Ed.), *International perspectives on maps and the Internet* (pp. 301–318). New York: Springer.

Turner, A. (2006). *Introduction to neogeography*. Sebastopol, CA: O'Reilly.

U.S. Department of Defense (2008). *GPS standard positioning service (SPS) performance standard*. Washington, DC: National Coordination Office for Space-Based Positioning, Navigation, and Timing Available at: http://www.gps.gov/technical/ps/2008-SPS-performance-standard.pdf.

United Nations. (2008). *United Nations environment programme: Indigenous knowledge in Africa*. Retrieved from http://www.unep.org/IK/

United States Geological Survey. (2016). *The National Map: Your source for topographic information*. Retrieved from http://nationalmap.gov/

Wilson, M. W. (2012). Location-based services, conspicuous mobility, and the location-aware future. *Geoforum, 43*(6), 1266–1275. doi:10.1016/j.geoforum.2012.03.014.

Wood, S. A., Guerry, A. D., Silver, J. M., & Lacayo, M. (2013). Using social media to quantify nature-based tourism and recreation. *Scientific Reports, 3*, 2976. doi:10.1038/srep02976 . Retrieved from http://www.nature.com/articles/srep02976#supplementary-information

You, M., Chen, C., Liu, H., & Lin, H. (2007). A usability evaluation of web map zoom and pan functions. *International Journal of Design, 1*(1), 15–25.

Zook, M. A., & Graham, M. (2007). Mapping DigiPlace: Geocoded Internet Data and the Representation of Place. *Environment and Planning B: Planning and Design, 34*(3), 466–482. doi:10.1068/b3311.

Zook, M., Shelton, T., Poorthus, A., Donohue, R., Wilson, M., Graham, M., & Stephens, M. (2015). What Would a Floating Sheep Map? In *The Manifesto* .Retrieved from http://www.floatingsheep.org

Chapter 7
Discovery and Fitness for Use

Abstract The purpose of this chapter is to provide readers with a succinct, but varied list of resources for obtaining geographic information (GI) for use, analyses, and geovisualization. We necessarily limit this list to authoritative outlets, such as those from federal, state, and local organizations, as well as private data vendors that are actively engaged in secondary data markets. To facilitate description, data types are subdivided into the 16 National Geospatial Data Asset (NGDA) themes identified by the Federal Geographic Data Committee (FGDC). Fitness for use, as a concept, is detailed and an example is provided using telecommunications data.

7.1 Geographic Information Discovery

Over the past several years, the terms "data discovery" and 'knowledge discovery" have become more mainstream and are now used in a variety of contexts. Data discovery refers to the process of identifying and obtaining observational data for use and analyses (Ames et al., 2012). Knowledge discovery is a more involved process that requires the application of data mining and analytic techniques (e.g., cluster analysis) to discover hidden and/or buried knowledge (Grubesic, Wei, & Murray, 2014; Zhu & Davidson, 2007) and interesting patterns in an obtained dataset. As detailed by Miller and Han (2009, p. 3), knowledge discovery must identify properties and "relationships that are valid, novel, useful, and understandable". For example, to be novel, the discovered patterns must be unique and/or unexpected in some way. To be valid, the discovered pattern cannot simply be an anomaly within the current database, instead the patterns need to be general enough to be applied to other databases. To be useful, the discovered knowledge must improve policy, decision-making, or strategy in a measurable way. Lastly, to be understandable, humans must have the ability to interpret, make sense, and reason about the discovered pattern (Fayyad, Piatetsky-Shapiro, Smyth, & Uthurusamy, 1996).

For the purposes of this chapter, we focus on data discovery. At first blush, data discovery may seem to be a relatively simple process—find data, perform analysis. However, given the widespread production and consumption of geographic information (GI), the discovery process is becoming more complex. There are a variety of challenges worth noting. First, much of the observational and/or environmental,

W. Bishop, T.H. Grubesic, *Geographic Information*, Springer Geography,
DOI 10.1007/978-3-319-22789-4_7

cultural, or socio-economic data published on the Internet is not easily discoverable. Although there are many open data efforts, such as the recent high profile offering from the city of Los Angeles (https://data.lacity.org/), many of these data are buried within databases that are not indexed by commercial search engines or crawlers designed to index the deep web. Second, as detailed by Ames et al. (2012), both *syntactic* and *semantic* heterogeneity in these data largely prevent their discovery, integration, and synthesis. For example, syntactic heterogeneity refers to the way data are encoded or organized. GI is available in hundreds of different encodings, including the relatively common shapefile (.shp) format, but also layer files (.lyr), Google Earth documents (.kml or .kmz), compact data format (.cdf), or many other encoding schemes. This can make it difficult for end-users to both determine the best format for acquisition and use, as well as the software platform to be used for analysis. Semantic heterogeneity refers to the identification of semantically related objects in different databases and the resolution of schematic differences among them (Kashyap & Sheth, 1997). For example, if the source of information and the potential receiver of that information operate within different contexts (Kashyap & Sheth, 1997; Madnick, 1999), the ontological terms and associated metadata for a database may differ dramatically. Further, the repositories of these data and their underlying structure may also diverge. In short, the process of data discovery is not always straightforward.

Fitness for use is also a concern for data discovery. Broadly defined, fitness for use is a multifaceted concept that refers to data suitability for a particular application or purpose. Facets include data quality, scale, interoperability, cost, metadata, syntactic and semantic heterogeneity, among others (Chrisman, 1984; de Bruin, Bregt, & Ven, 2001; Veregin, 1999). For example, although the visible satellite imagery provided by the WorldCat 3 satellite may be highly suitable for a project that requires a precise situational awareness, obtaining this imagery may be cost prohibitive. As a result, the fitness for use of the WorldCat 3 imagery may be less than ideal.

Perhaps the biggest challenge facing researchers engaged in geospatial data discovery is the sheer volume of data available for acquisition and analysis. Geospatial data generated via social media platforms (e.g., Twitter), unmanned aerial systems, satellites, sensor networks, and mobile devices using global positioning systems (GPS) are major contributors to the 2.5 quintillion bytes of data generated globally each day (Walker, 2015). Within this mix, federal, state, and local agencies, as well as private data vendors, both in the U.S. and elsewhere, are creating big geospatial data relating to climate, soils, transportation, demography, and many other domains that are discoverable.

In an effort to help make sense of geospatial data discovery and its fitness for use, this chapter is structured to provide readers with a curated list of resources for obtaining GI for analysis and geovisualization. This list is not comprehensive, nor does it represent a complete overview of the data available to researchers. Because of space constraints, we limit this list to the more traditional, authoritative outlets, such as those from federal, state, and local organizations, as well as private data vendors that are actively engaged in secondary data markets.

We openly acknowledge that volunteered geographic information (VGI) (Sui, Elwood, & Goodchild, 2013) and much of the content found on the Geoweb (Chap. 6; Bishop, Grubesic, & Prasertong, 2013) is of growing importance and may be equally discoverable. However, for the purposes of this chapter, we are forced to narrow the field somewhat. In an effort to facilitate description, we draw upon the 16 National Geospatial Data Asset (NGDA) themes identified by the Federal Geographic Data Committee (FGDC). Fitness for use, as a concept, is also detailed and an example is provided using telecommunications data.

7.2 Public Domain Data and the National Geospatial Data Asset (NGDA) Themes

As detailed by Kerski and Clark (2012), there is no universally accepted definition for public domain, but it is widely accepted that it adheres to some basic themes, such as lack of copyrights, freedom from intellectual property rights, no requirement to pay royalties, and open source development principles. That said, data that are publicly available, do not always conform to the principles of the public domain. For example, Google Maps are freely available for all to use but they cannot be redistributed, published, copied, modified, or used in any derivation without prior written authorization from Google (2016). In other words, these data are publicly available, but not part of the public domain.

The good news is that there are petabytes of public domain data that can be used for spatial analysis and geovisualization with no restrictions or copyright constraints. In an effort to better organize and document a portion of these data, the Executive Office of the President (2010) released Memorandum M-11-03, "Issuance of OMB Circular A-16 Supplemental Guidance", with the goal of improving the implementation of "OMB Circular A-16, Coordination of Information and Related Spatial Data Activities". The overarching point of this effort was to create an enhanced data management process that would provide federal agencies and other stakeholders with a stronger framework for managing spatial data portfolios. This is important for federal agencies because the ability to efficiently share, use, and invest in geospatial assets is critical to the day-to-day operations of agencies, governments (at all levels, federal, state, and local) as well as other stakeholders with an interest in GI. In particular, the NGDA is structured to accomplish the following (FGDC, 2014, p. 3):

- Increase the opportunity for shared use;
- Increase the value of data as more missions and partners use and rely upon it;
- Increase the opportunities for partnering in the data's creation, development, and maintenance;
- Quantifies data investment against its return (i.e., how much data for how much cost); and
- Provides mechanisms for identifying data gaps and investment requirements.

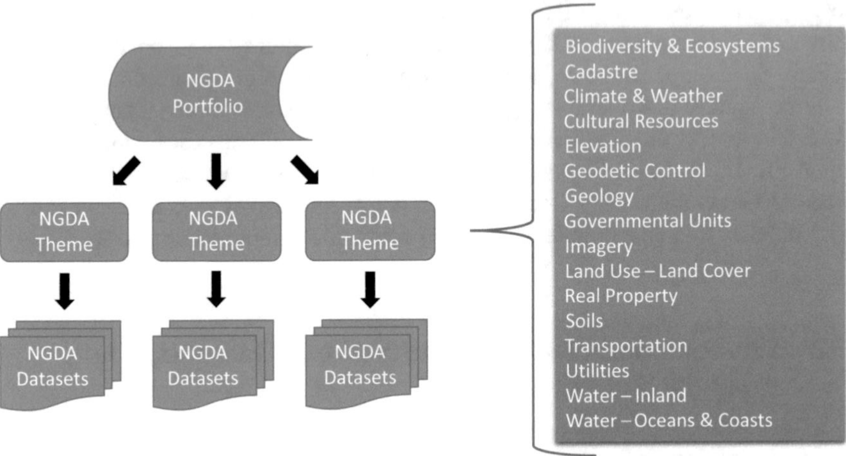

Fig. 7.1. Overview of the NGDA theme organizational structure

One of the outcomes associated with this effort is a large suite of discoverable GI, available for download and analysis. Figure 7.1 provides a basic overview of the 16 NGDA themes and their organizational structure. As of mid-2016, all of these data were stored and available to the public at: http://www.fgdc.gov/ngda-reports/NGDA_Datasets.html. It is important for readers to understand that the NGDA datasets do not represent the complete universe of public domain geospatial data at the federal level. However, according to the FGDC (2014), the data from FGDC members do represent a large portion of the federal geospatial data used across multiple programs, agencies, and partner organizations. In total, there are sixteen NGDA themes and all contextual descriptions are current as of 2016.

7.2.1 Biodiversity and Ecosystems

The biodiversity and ecosystems theme includes seven datasets listed via data.gov, four of which concern ecoregions and the remaining three pertain to species distributions and ranges. Both Bailey's (1980) and Omernik's (1987) ecoregions are included in two separate sets, as well as an Environmental Sensitivity Index dataset. Two datasets of species ranges and distributions that came from the U.S. Geological Survey (USGS) Gap Analysis Program are also included, but are only available at a fairly coarse resolution. Finally, the Fish and Wildlife Service's Critical Habitat for Threatened and Endangered Species dataset is included, which is often used in conservation publications, as well as technical publications produced by U.S. government agencies. These datasets are very large, and all seven include up-to-date metadata.

This theme's list of available data will continue to expand. Many of the biocollections located in the United States that have been housed in natural history museums, herbaria, private collections, zoos, and so forth, are getting digitized and curated in formal repositories. In addition, non-FGDC members have plenty of high-quality GI for this theme. For example, the National Science Foundation's (NSF) Advancing Digitization of Biodiversity Collections (ADBC) addressed digitization through the iDigBio project (https://www.idigbio.org/). iDigBio's mission was to develop a national infrastructure of standards and best practices for digitizing biocollections and as of mid-2016 the iDigBio portal, contains 60,839,108 specimen records and 14,173,310 media records. Globally there is a near limitless number and variety of biota from deep-sea fish to tundra fungi to fossilized trilobites. The amount and diversity of biodiversity and ecosystems theme users also makes this a valuable, but complicated theme to navigate for discovery and fitness for use (Bishop & Hank, 2016). Biodiversity and ecosystems theme users include decision-makers and industry professionals related to agriculture, food security, public health, genomics, bioprospecting, ecotourism, mining, forestry, as well as the educators and students from elementary through graduate education. Each group has differing fitness for use considerations whether studying evolution, species poleward migrations, or teaching primary schoolers about biodiversity.

7.2.2 *Cadastre*

This is one of the NGDA framework themes and includes GI used to describe property extents and the entities that hold jurisdiction over them. There are over 20 unique cadastral datasets catalogued by the NGDA, ranging from the Bureau of Indian Affairs Indian Lands extents to National Parks extents. Also included are several continental shelf lease block datasets, commonly used for energy research, especially offshore drilling and wind sites. Finally, public land survey data that includes parcel data and parcel attributes are included. Notably missing are some federal and military facility footprint data. While Army land tracts are included, Department of Defense parcels are not. Cadastre are comprehensive real estate lists; therefore, this theme globally would include a mindboggling number of datasets for all land parcels, with accompanying values, uses, and ownership history. Potentially any user operating in the built environment needs access to these GI, but especially those working in urban planning, constructions, real estate, and every level of government administrating land use.

7.2.3 *Climate and Weather*

This theme is led primarily by the National Oceanic and Atmospheric Administration (NOAA), and includes meteorological conditions in a particular region over time. The bulk of these data are remotely sensed weather data over space and time and include databases on sea surface temperature, for example. Other databases include radiometer data from polar-orbiting satellites, real-time precipitation, and radar data from NEXRAD, and elevation data from the PRISM dataset. Most of these include time series from whenever the data were initially being remotely sensed. The sea surface temperature data, for example, starts in 1981 and stretches to the present. Each time you check the weather on a mobile device and decide whether or not to take an umbrella, you are accessing data from this GI theme. The National Weather Service (NWS) data, as well as terabytes of raw GI coming from countless surface stations and satellites around the globe are leveraged for both public safety and commercial purposes. Earth and climate scientists use these data for many purposes—from measuring aerosol pollution to improving forecasting algorithms for severe weather conditions.

7.2.4 Cultural Resources

Led by the National Park Service, these datasets are designed to show features of places that are historically, culturally, or architecturally significant to society, but often include important feats of engineering. There are only two major products associated with this theme—a shapefile and attributes of the National Register of Historic Places and the Geographic Names Information System (GNIS), which details federal standards for nomenclature across all government applications. The National Register of Historic Places serves as a useful dataset for anyone working in the geohumanities, especially when attempting to discover places of import over time. Similar datasets exist in other countries as well as some global scale products that fit this theme (e.g., United Nations Educational, Scientific, and Cultural Organization (UNESCO) World Heritage sites polygon http://whc.unesco.org/en/news/853/).

The GNIS was developed in cooperation with the U.S. Board on Geographic Names (USBGN) to supervise placenames standardization (i.e., each named feature should have only a single authorized name). Globably this theme would also include other country-specific agencies and supranational groups such as the United Nations Group of Experts on Geographical Names (UNGEGN) to assure that each placename used matches a single locality.

7.2.5 Elevation

This is another of the framework themes and includes datasets designed to describe natural surface elevation, bathymetric data, building heights, and derived data such as slopes, aspects, and reliefs. There are over ten elevation datasets in this group, including the widely used USGS Digital Elevation Models (DEM) (at various resolutions), but also newer products that provide radar topography data. A handful of

Light Detection and Ranging (LiDAR) datasets are also available, largely focusing on coastal areas. Finally, there is a FEMA flood hazard shapefile for the entire country that is used in hazard modeling research and by local agencies in designing hazard evacuation and mitigation plans.

7.2.6 Geodetic Control

Geodetic control data, provided by NOAA, includes a handful of technical datasets used to help maintain geospatial consistency, with datasets tied to the National Spatial Reference System. There are active and passive geodetic control shapefiles, a dataset of several geoid models that can be used for earth surface modeling, and there is a gravity dataset covering airborne gravity of the U.S. dating from 2008. These shapefiles can be used for several models and, in a few instances, to build GPS functionality into web mapping applications.

7.2.7 Geology

There are only six datasets included in this theme, three of which are shapefiles of oil and natural gas wells in a variety of places (Pacific Region, Alaska, and the Gulf of Mexico). Another is a shapefile of the sand and gravel borrow areas that the federal government has access to, while there is also a seismic anomaly shapefile for the Gulf of Mexico. The sixth is an archive of stratigraphic information and historical maps maintained by the National Geologic Map Database—over 90,000 maps are included beginning from the year 1800. Notably absent are datasets including other geological features, such as paleontology, hazards, karst, and earth processes.

7.2.8 Government Units, and Administrative and Statistical Boundaries

This theme houses all of the U.S. Census datasets. There are dozens of separate datasets here, ranging from the U.S./Canada border shapefile to U.S. County subdivisions and the 2010 Census population counts. These datasets are used for a wide variety of purposes, from academic research to community planning and policy generation. There are also several datasets used for emergency planning such as school district boundaries and severe weather warning area coverages. Although not listed in data.gov, the Minnesota Population Center (MPC) provides free online access to many historical and international census and survey data, including the Integrated Public Use Microdata Series (IPUMS) (usa.ipums.org), National Historical Geographic Information System (NHGIS) (nhgis.org), and Terra Populus (terrapop.org).

7.2.9 Imagery

Datasets within this framework theme are collections of remotely sensed orthoimagery. The datasets are commonly used for research, industry, and government as base maps for spatial analysis. The datasets include Landsat imagery dating back to Landsat 1 (1972), MODIS (both terrestrial and aquatic), and ASTER for thermal radiation. There are also several datasets specific to certain regions or themes—these include a NOAA coastal mapping remotely sensed dataset, and a National Agriculture Imagery Program collection of agriculturally relevant images. Ultimately, the NGDA determines what is listed under this theme through their platform, but available imagery is quite extensive, potentially comprising a great deal of remotely sensed GI and a large variety of users beyond the FGDC members populating the current GeoPlatform. For example, the USGS Earth Resources Observation and Science (EROS) Center serves up many web-based tools to make imagery more accessible, including TerraLook (https://lta.cr.usgs.gov/terralook/home).

7.2.10 Land Use–Land Cover

Datasets here are often built from datasets in the *Imagery* theme, but pertain to specific issues of interest. For example, there are coastal change datasets, cropland maps, historical fire maps, wildfire trends and potential severity, and the National Land Cover Database collection of land cover maps and images. These datasets are used for historical land cover change research, to study various hazards, and for agricultural planning and zoning. This group also includes the North American Land Change Monitoring System (NALCMS), which is a trilateral effort between the U.S., Mexico, and Canada to develop a harmonized, highly accurate, and consistent multi-scale land cover monitoring data-stream that captures changes throughout North America. The USGS Land Cover Institute (LCI) serves GI from this theme from local to global scales, but this and many other relevant datasets do not currently appear in the NGDA platform.

7.2.11 Real Property

Many of these datasets are similar to the cadastral datasets, but with additional ancillary attribute information. The theme is designed to house datasets that show the spatial location of real property units, along with generic use attributes. Datasets include shapefiles of federally owned properties, including assisted housing units, Federal Housing Administration properties that are for sale, properties owned or leased by the General Services Administration, and public housing buildings. These datasets are most commonly used for government administrative efforts, but in concert with additional real property datasets not owned federally, they will be increasingly adopted for use in the socio-economic and planning sciences.

7.2.12 Soils

The dataset most iconic to this theme is the U.S. General Soil Map, displaying surface soil information for the entire U.S. There are also databases that detail soil horizons, to whatever depth has been surveyed for a given region. The primary government agency in charge of maintaining these datasets is the Natural Resources Conservation Service. These datasets are often used to reconstruct past geologic events, and have recently been used as an aid in predicting plant range shifts in light of future climate change.

7.2.13 Transportation

Datasets within this framework theme fall into two main categories. The first is transportation network data, such as the commonly used Topographically Integrated Geographic Encoding and Referencing (TIGER) street network for the U.S. Other transportation network data include waterway lines and railroad lines. The second category is ancillary transportation data. These include point geographic basefiles of airports, bridges, locks, and more and navigational charts for inland waterways. There are also products dealing with traffic counts and automobile accidents. The majority of the data within this theme are used for planning efforts (e.g., traffic flow, hazard evacuation plans, access and accessibility, etc.), as well as economic analyses dealing with the transport of time-sensitive goods.

7.2.14 Utilities

Currently, only two datasets are maintained in this theme—oil pipelines and oil platforms in the Gulf of Mexico. These are used for oil spill modeling, future energy location surveys, and hurricane hazard mitigation planning. The theme is also designed to include datasets dealing with other utilities such as water treatment, drinking water distribution, and communication. The Bureau of Safety and Environmental Enforcement is the lead government body in charge of the offshore utilities datasets but the terrestrial utilities have no such managing agency yet, hence their absence from the theme.

7.2.15 Water–Inland

Inland water datasets are managed by the USGS and the U.S. Fish and Wildfire Service. This theme includes interior hydrography datasets of the entire U.S., a national dam inventory, a national levee database, and a national wetland inventory. There is also a geographic base file of the boundaries of watersheds within the country. These datasets are used for dam planning, treaty negotiation, and interior flooding risk mapping. Currently absent but within the theme's scope are datasets dealing with groundwater information, navigable waterways, and water quality information.

7.2.16 Water–Oceans and Coasts

Datasets in this theme range from the technical water boundaries of the U.S. to raster–based navigational charts used for open water vessels. There are also shapefiles of revenue zone boundaries between the U. S. and its neighbors as well as observations of natural features such as sea level variations within the boundaries of the U.S. These datasets are used for negotiation, determining fishing rights, as actual navigational aids, and for some marine conservation research (specifically a dataset on the U.S. marine protected area boundaries).

7.2.17 Summary

There is no doubt that the detailed NGDA data themes cover a wide range of data collected by federal agencies, from the geophysical to the socio-economic and demographic. Of note, GI concerning the polar regions of the planet typically have separate, specific portals for discovery of that geospatial data. The NSF-funded Polar Earth Observing Network (POLENET) (polenet.org) and several web-based mapping applications from the Arctic Institute of North America (arcticconnect.org) provide access to the GI for the poles. Overlap between themes is inevitable, but the NGDA themes identified by FGDC can help organize GI discovery across the various creators and communities of users beyond government GI. Although many of these data can be used "as is" for spatial analysis and visualization, over the past several decades, an entire industry has emerged that uses these data as the backbone for enhanced, secondary data products. In the next section, we provide readers with a brief glance into the basics of this process and discuss the discoverability of these data for analyses and visualization purposes.

7.3 Enhanced Data Discovery

The idea of enhanced data may seem somewhat ambiguous to readers. For example, many data can be enhanced through basic analytics (e.g., spatial aggregation), making them more meaningful or easier to use. In other instances, data can be enhanced through abstraction (e.g., conversion to points, lines, and areas), making them simpler and more easily interpretable. In short, enhancements can manifest in many different ways, but the key process is one which takes primary data and adds value (in some way) for the benefit of the end-user.

Consider, for example, the TIGER/Line files made available from the U.S. Census Bureau (2015). TIGER consists of a digital database of geographic features such as administrative political units (e.g. counties), statistical boundaries (e.g., block groups), infrastructure (e.g., roads, rail), and geophysical features (e.g. rivers and lakes) that covers the entire U.S. Easily discoverable, these data are used for a variety of mapping and analysis tasks. However, in their primary form, TIGER/Line files do not have any socio-economic or demographic information tied to the statistical boundaries, such as tracts. Instead, the TIGER/Line files contain geographic entity codes that serve as unique identifiers for linking Census demographics to each observational unit in the database. For example, Fig. 7.2a illustrates the basic features of the TIGER/Line database—streets, administrative boundaries, tracts, and hydrographic features. Figure 7.2b displays the same basemap, but with linked socioeconomic information from the Census. In this case, 7.2b illustrates 2012 median household income estimates, by tract. Although the raw TIGER/Line files are easily linked to Census data, many end-users prefer to avoid this somewhat time-consuming process and receive the data in post-processed form. In turn, many secondary data vendors are happy to provide these products. For example, Environmental Systems Research Institute (Esri) supplies many different types of enhanced TIGER/Line and Census demographic data products (ESRI, 2015). In addition to the American Community Survey (ACS) data, which are pre-linked to a range of TIGER/Line statistical boundaries (e.g., block groups, tracts, counties), Esri also makes available a suite of updated demographics which attempt to fill in the gaps between Census surveys. Where the former is concerned, the added value is strictly convenience-based. The data are ready to use and there is no need for the end-user to obtain information from the American Factfinder site for analyses. For the latter, the same rules apply, but the data estimates are more timely and there are a range of additional variables (not reported in the Census) that Esri can provide, including consumer preference data, spending habits, and other geodemographic indicators. Esri is relatively transparent regarding its demographic update methodology, publishing it for public inspection (http://tinyurl.com/j22kctf).

In either of the above instances, the data are easily discoverable, but there are concerns regarding fitness for use (Graves & Gerney, 2016) that will be detailed later in this chapter. Also, it is important to note that Esri is not the only provider of these types of pre-processed data, companies such as Alteryx (http://www.alteryx.com/),

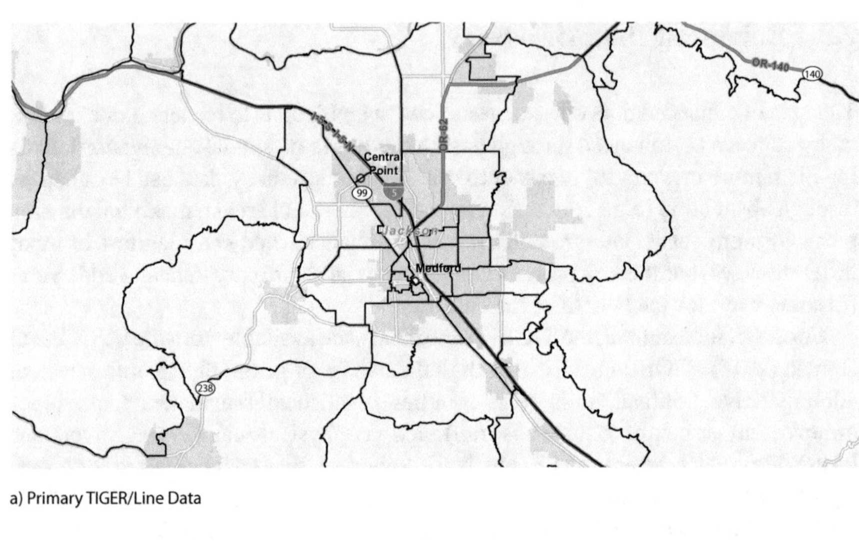

a) Primary TIGER/Line Data

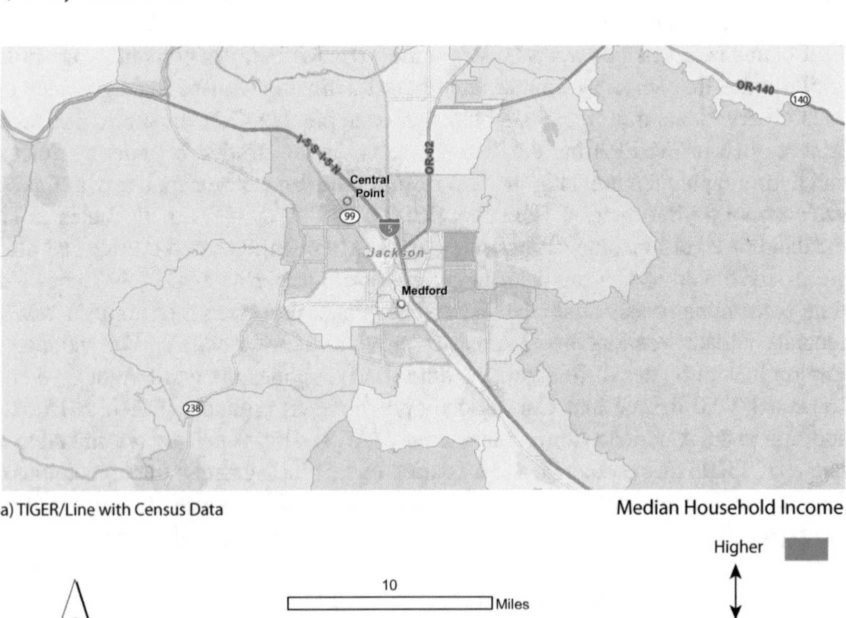

a) TIGER/Line with Census Data

Fig. 7.2. Fusion of TIGER/Line Data and Census Demographic Data

Synergos (http://www.synergos-tech.com/), Geolytics (http://www.geolytics.com), as well as Caliper (http://www.caliper.com) and Pitney Bowes (http://www.pitney-bowes.com) all provide enhanced spatial and demographic data for use.

Enhanced data are not limited to demographic and socio-economic databases that are derived from the Census. Enhanced data are discoverable for nearly every domain of inquiry. Listed below are a handful of examples for domain specific applications that may be of interest to readers.

7.3.1 Here (Navtech) Streets

Here Global, formally known as Navtech, is owned by a consortium of automakers (the VW Group, BMW, and Daimler). Here provides a wide-range of mapping data and navigation-based technologies for the automotive, direct-to-consumer, and enterprise markets. One of the more notable offerings is the Here Map Content for street networks. For example, in the U.S., in addition to using the basic TIGER/Line content (found in the Here base package), the premium version of the street network also includes information regarding directional restrictions (e.g., one way), historical traffic flow, a massive array of contextual content such as restaurants, entertainment venues, business facilities, and educational institutions. In places where road information is not available, or there are questions concerning the overall accuracy of the acquired source material, Here dispatches field crews that drive the streets to collect and verify base content. Field crews also log details on navigational obstructions and limitations for their Trucks product, such as vehicle height restrictions, weight limits, and turn restrictions. Lastly, Here leverages high-resolution visual satellite imagery for verification purposes and operates a user platform where millions of daily updates are transmitted (https://mapcreator.here.com/mapcreator/) to help improve the street network.

7.3.2 Pitney Bowes GeoEnrichment

One of the major challenges in spatial analytical work is connecting street addresses to their real-world locations (Goldberg, 2011). In the U.S., the majority of commercial geocoding engines use address information, TIGER/Line data, and a suite of proprietary interpolation techniques to assign latitude and longitude coordinates to each street address. Output quality is often questionable (Cayo & Talbot, 2003; Ratcliffe, 2001), which forces users to pursue alternative approaches (Murray, Grubesic, Wei, & Mack, 2011). The GeoEnrichment module from Pitney Bowes uses a master address fabric that contains every known address in the U.S. (170+ million) and their associated parcels (i.e., cadastral data) to create a set of persistent geographic coordinates. This includes millions of non-postal street addresses within gated communities. Coverage issues likely remain in rapidly growing areas, but the use of persistent coordinates that are not solely generated from TIGER/Line interpolations represents a significant leap in locational quality.

7.3.3 Telecommunications Data

There are many easily-discoverable telecommunications products for spatial analysis and geovisualization. For example, the National Broadband Map (NBM) (http://www.broadbandmap.gov/) in the U.S. provides consumers (and researchers) with block-level information on the number of providers offering service within each block, advertised speeds, and a number of other metrics. The NBM is no longer being updated and the June 2014 data collection is the last one to be maintained by the National Telecommunications and Information Administration (NTIA). The most current information available on broadband in the U.S. is once again the Federal Communication Commission (FCC) Form 477 data (http://tinyurl.com/z43ydp3). The FCC also provide basic information on cellular telecommunications towers through its antenna structure registration system (http://tinyurl.com/hjkqhr5). Additional telecommunications data, such as wire-center boundaries (Grubesic & Murray, 2005; Grubesic, Matisziw, & Murray, 2011, local access transport areas (LATA) (Sherali, Lee, & Park, 2000), and public safety answering points (PSAPs) (Campbell, Gridley, & Muelleman, 1997) are available from a range of providers including Pitney Bowes, Here, and Geo-Tel (http://www.geo-tel.com/), among others.

7.3.4 Business Data

There are several different sources for business data in the U.S. that offer nation-wide coverage. Within the public domain, Economic Census (http://www.census.gov/econ/census/) is the official measure of American business and the economy. Industries are categorized by the North American Industry Classification System (NAICS) (e.g., 311411 = frozen fruit and juice manufacturing; 541360 = geophysical surveying and mapping services). The American Factfinder from the Census Bureau provides access to these data in the form of county and ZIP code business pattern data. All of these data are in the public domain. However, there are also a number of enhanced business databases that are available for analysis, including those provided by Dun and Bradstreet (D&B) and InfoUSA Inc. The InfoUSA (2016) data currently include 16.1 million verified business records that are cross-checked with information from other sources such as federal and state business registries, phone verification, and information from the U.S. Postal Service. The Esri Business Analyst (ESRI, 2016) is a popular portal to these data, providing a ready-to-use shapefile of all 16+ million geolocated businesses, sorted by NAICS codes. Previous validation work suggests that the InfoUSA data made available in Business Analyst includes approximately 51% of all business types (Hoehner & Schootman, 2010). This means that the local business environment reflected by the InfoUSA data is a conservative representation, at best. In contrast, the Dun & Bradstreet (D&B) business data reflect their efforts to develop predictive credit ratings and scores for use in the banking and insurance industry. As a result, businesses have an incentive to respond to the D&B survey efforts (Kaufman et al., 2015) and the resulting D&B data are relatively comprehensive, including information on NAICS codes, number of employees, and a variety of other descriptive characteristics (Bader, Ailshire, Morenoff, & House, 2010; Boone, Gordon-Larsen, Stewart, & Popkin, 2008). One derivative of the D&B data is the National Longitudinal Establishment Time-Series (NETS) database from Walls and Associates (Walls,

2007). NETS includes approximately 300 variables per business, including geographic coordinates (latitude and longitude), relocation history, NAICS codes, and the number of employees, among many other indicators.

7.4 Fitness for Use

The core principle of *fitness for use* is relatively simple. Chrisman (1984, p. 81) suggests that "the foundation of data quality is to communicate information from the producer to a user so that the user can make an informed judgement on the fitness of the data for a particular use." Although this seems reasonably easy to accomplish, communication paths (producer → user) are rarely direct, semantics vary, there are temporal lags between data creation and use, and the technical expertise of users varies significantly, potentially diminishing their ability to make an informed evaluation of the available data. In addition, as the discipline of GIScience continues to mature, the inherent complexities of the problems tackled in this domain have increased. While rudimentary proximity analysis and distance calculations remain important, analytical tasks that include complex spatio-temporal interactions and big geospatial data are propelling the need for improved spatial analytical tools and data quality. Where the latter is concerned, the persistent and pervasive lack of data documentation diminishes the ability for analysts/users to determine fitness for use. As detailed in Chap. 5, although the development of metadata standards over the past 20 years (e.g., FGDC, ISO, ANZLIC, and so forth) has helped, the adoption of these standards across agencies and producers/users in the field is inconsistent.

As mentioned earlier, fitness for use is not simply a measure of data quality, nor does it strictly focus on the absence of errors in the data (Morrison, 1995). Specifically, internal data quality usually consists of measures or metrics that account for positional accuracy, temporal accuracy, attribute accuracy, logical consistency, and completeness (Guptill & Morrison, 1995). In most cases, this information is transmitted from the producer to the user via metadata. The problem with metadata is that many users choose to ignore it (even when metadata are present). There are many reasons for this, ranging from the metadata being too complicated or difficult to interpret (Devillers, Bédard, Jeansoulin, & Moulin, 2007), to the metadata being housed separately from the data they detail, making it challenging to explore quality information directly from GIS graphical interfaces. Again, metadata and its associated problems and prospects are discussed thoroughly in Chap. 5.

Fitness for use is a more holistic concept and evaluation framework that includes the core elements of internal data quality (e.g., positional accuracy, temporal accuracy, and so forth), but also includes a varied suite of supplemental indicators that help analysts and decision-makers to evaluate the context and implications of data use for a particular application. For example, we know that airline seat inventory control is a complex process, largely designed to balance and/or optimize the number of discount and full-fare reservations for any given flight (Belobaba, 1987). The end-game, of course, is to maximize revenues for the airline. If too many discount fares are offered, passenger loads increase, but yields decrease. The reverse can also occur, when too many business fares are sold and planes are flying half empty because price-conscious travelers rejected higher fares for their travel plans. The management of fare price points, seat inventory, and the timing of fare offerings is highly choreographed, and there are many data points and associated variables created during the process. However, two simple indicators of the relative success or failure of the management process include "passenger load factor" and "passenger yield". Passenger load factor is the ratio of passengers miles traveled to seat miles available. As values trend toward 1, it indicates that planes are operating near capacity. In short, it is a measure of the proportion of the airline output that is consumed by passengers. Passenger yield is a measure of the average fare paid per mile, per passenger. As passenger yields increase, airline revenues are improved. When combined, passenger load factor and passenger yield generate a third indicator, *passenger revenue per available seat mile* (PRASM) that can be used to compare operational goals across routes for an airline.

Regardless of the indicator used, the point of this example is to suggest that analysts know that airline operations are complex and that they cannot possibly observe every nuanced characteristic when evaluating airline operations or performance. As a result, a subset of measures are used to gain a broad view of operational health. If one of the indicators is problematic (e.g., low load factors), additional analytics to identify the root cause of the problem may be required. Klein (1999) notes that many analysts, decision-makers, and high-level professionals are required to make important decisions within relatively short time-frames. The example indicators detailed above can serve as cues in this process, at least for the airline industry, and ultimately drive the logic behind important, time-constrained decision-making. The use of indicators is unavoidable when dealing with large volumes of information, even if it is widely recognized that indicators are imperfect. In addition, personal experience and professional training also factor into this process. In sum, there is no single recipe for determining data fitness for use. It is context dependent and conditional upon the application domain. For readers interested in learning more about fitness for use, or interactive and/or automated systems that help determine fitness for use, see (de Bruin et al., 2001; Devillers, Bédard, & Jeansoulin, 2005; Devillers et al., 2007; Heipke, 2010).

7.4.1 Fitness for Use: The National Broadband Map

The NBM was a joint effort between the FCC and the NTIA. It was motivated by the Broadband Data Improvement Act (BDIA) of 2008, which called for efforts to improve the quality of state and federal broadband data. In particular, the BDIA sought to improve available information concerning *where* broadband was offered, as well as measures concerning quality of service. In July 2009, the Broadband Technology Opportunities Program (BTOP), which was funded as part of the American Recovery and Reinvestment Act of 2009 (Kruger, 2009), released funds to begin the process of mapping broadband in the U.S. As detailed by Grubesic and Mack (2015), in an effort to maintain consistency and regularity in the data collection efforts, each state selected one entity for acquiring, cleaning, and tabulating broadband data from local providers and ultimately injecting these data into the NBM. For example, in the state of Oregon, the Oregon Public Utilities Commission (OPUC) was charged with collecting broadband data, but ultimately the OPUC selected a subcontractor to execute the mission.

Nearly simultaneous to the NBM efforts, the FCC released the national broadband plan (*2010 Plan*) for the U.S. (FCC, 2010). As detailed by Grubesic (2012a), the *2010 Plan* identified broadband as a crucial technological pathway—one that could improve health care, education, energy use, economic opportunity, government performance, civic engagement, and public safety through a more fluid and dynamic exchange of information and knowledge via broadband networks. The resulting FCC strategy provided a multifaceted agenda for developing and enhancing broadband infrastructure throughout the U.S.

It was hoped that when the NBM and its data were combined with the conceptual components of the *2010 Plan*, residents, businesses, local, state, and federal government agencies, and all entities that had a vested interest in advancing regional development efforts would be able to develop a better understanding of broadband provision, access, use, and its related economic, and social impacts. The results of these efforts were mixed, at best. To be sure, the NBM data provided fuel for exploring adoption gaps (Whitacre, Strover, & Gallardo, 2015), broadband's impact on employment (Jayakar & Park, 2013), and the relative successes/failures of the BTOP awards that were allocated (LaRose et al., 2014); however, the NBM data and their fitness for use across a wide variety of applications were strongly questioned (Grubesic, 2012a, 2012b).

One pertinent example of NBM data and its fitness for use pertains to efforts for evaluating the number of households provided coverage with digital subscriber line services (xDSL) throughout the U.S. xDSL is a copper-based broadband platform that allows for the overlay of a high-capacity data channel on top of a standard analog voice channel on regular telephone lines (Grubesic & Mack, 2015). Although accurate metrics regarding broadband market share are difficult to obtain, recent estimates suggest that xDSL continues to thrive in the U.S. (Elliott, 2014), with providers such as AT&T, Verizon, and CenturyLink gaining market share against the larger cable providers such as Comcast, Time Warner, and Cox. Unfortunately, broadband providers do not make their subscriber data available for inspection, which means that efforts to understand where these broadband services are offered are forced to use aggregate data, such as that from the NBM.

The problem with xDSL coverage data, at least as reported by the NBM, is that each state entity charged with developing a methodology for calculating xDSL service coverage used different standards (Grubesic & Mack, 2015). For example, in Massachusetts, the coverage standard was a network distance of 17,800 ft. from the xDSL switch, with an additional 200 ft. buffer to account for the distance limits of most asymmetric DSL services. However, in Illinois, Euclidean (i.e., straight line) buffers of varying sized were generated around each xDSL switch and clipped to ensure that the service area estimates did not exceed telephone wirecenter service areas. In West Virginia, xDSL coverage was estimated with 15,000 ft. service areas. Worse, virtually no information regarding the geocomputational procedures used to flag Census blocks as "served" or "unserved" were provided at the state level, let alone with the final data reported by the NBM. Complicating matters is that household counts and their associated tabulations for subcounty scale demographic data are also relatively uncertain. In recent work, Graves and Gerney (2016) demonstrated that, at a minimum, household counts provided by secondary (i.e. enhanced/value added) data vendors vary in suburban North Carolina by more than 10% in best case circumstances and up to 30% or more for 5-year population projections. Readers should note that these differences likely represent the most conservative estimate of potential problems with these demographic data. When subjected to spatial aggregation or related tabulation procedures, local, and regional uncertainty is sure to increase. For more details on these issues and the problems with NBM data, see Grubesic and Mack (2015).

Where fitness for use is concerned, there are three major failings associated with the NBM and using its data for estimating the number of households covered by xDSL service. First, there is a lack of internal data quality within the NBM. As detailed above, states used different approaches for deriving and evaluating the relationships between xDSL coverage polygons and the block groups they potentially serve. Positional accuracy, thematic accuracy, logical consistence, and completeness (Morrison, 1995) are all a concern within this context. Second, although the NBM provides standardized metadata for each state and its broadband provision information, these data fail to acknowledge the inconsistencies in the data generation process for each state in a transparent way. Again, the geocomputational approaches used for calculating xDSL coverage are buried within methodology statements provided by each state—if this information is provided at all. Further, because this material is technically dense, it is unlikely that casual users will read or understand the content. Instead, most analysts will simply use the data, as they are. Lastly, there are massive data asymmetries within the NBM that are almost completely unacknowledged by the broadband community. Namely, several iterations of the NBM use mixed Census block geometries (i.e., both 2000 and 2010) for data collection and statistical tabulation. This is a huge problem because in 2000, there were 8,262,363 unique Census blocks, but this increased to 11,155,486 in 2010. This means that a direct comparison between these geometries is not possible, yet they exist simultaneously in the June and December 2011 iterations of the NBM. When combined with the household/population uncertainty highlighted by Graves and Gerney (2016), the overall fitness for use of data in the National Broadband Map is questionable, at best, for local and regional development initiatives.

Although this example presents numerous issues of error propagation within a national dataset, all GI requires similar considerations for users when determining fitness for use. For many, research purposes and potential spatial analysis are restricted by limitations in GI. For example, the Institute of Museum and Library Services (IMLS) Public Library (Public Use) Data Files have limited fitness for use for spatial analysis because library use data is aggregated to system levels eliminating the potential for outlet-level analysis, unique identifiers are recycled when closures and openings occur, and library outlets locations are inaccurately geocoded (Koontz & Jue, 2006). In other words, fitness for use is dependent on factors related to the original data collection and details outlined in accompanying metadata.

7.5 Conclusion

There is certainly no shortage of opportunities for geographic data discovery in the information, socio-economic, planning, environmental, and physical sciences. Open data portals continue to emerge, increasing the transparency of local, regional, and federal governments operations (Janssen, Charalabidis, & Zuiderwijk, 2012; Kitchin, 2014) and social networks are increasingly providing more data points concerning our preferences, habits, and daily activities (Lewis, Kaufman, Gonzalez, Wimmer, & Christakis, 2008; Wang, Yu, & Wei, 2012). Not surprisingly, we anticipate that GI availability will continue to grow exponentially into the foreseeable future, especially as the Internet of Things (IoT) begins to take shape and the Geoweb expands its reach. Specifically, as more devices become connected to the Internet, including thermostats, refrigerators, door locks and lighting systems, the discoverability of these data will certainly increase, providing an amazing window into the everyday rhythms of the planet, some easily seen, others not.

The purpose of this chapter was to provide readers with a succinct, but varied list of resources for obtaining geographic information for use, analysis, and geovisualization. We necessarily limited this list to authoritative outlets, including those from federal, state, and local organizations, as well as private data vendors that are actively engaged in secondary data markets, but readily admit that data from the GeoWeb and other, more ad hoc sources are of growing importance and value. Details regarding the 16 NGDA themes identified by the FGDC were provided, as were a range of enhanced data options, many of which build upon the federal data outlined in the NGDA themes.

Tying all of this material together was a discussion on data *fitness for use*. Although the core tenet of this process pertains to the ability of data producers to communicate information about the data to consumers/users so that they can make an informed judgement on the fitness for the data for a particular application, the evaluation process is rarely straightforward. Fitness for use judgements and frameworks tend to be complex, reflecting both the growing variety and volume of data now available, but also the need for more integrative and multifaceted analyses to draw knowledge from these data. Metadata is central to this process, but so too are professional experience and the ability to develop indices or cues to facilitate decision-making and fitness for use judgements in time constrained environments. Examples of fitness for use were illustrated using data from the National Broadband Map.

In sum, data discovery will continue to play an important role in GI organization, access, and use. Although the processes for data discovery will continue to evolve, so too will the data themselves. As a result, it is critical for producers, users, and information professionals to continue their efforts to document, provide context, and clarify the semantic structure of these GI to ensure portability and reuse.

References

Ames, D. P., Horsburgh, J. S., Cao, Y., Kadlec, J., Whiteaker, T., & Valentine, D. (2012). HydroDesktop: Web services-based software for hydrologic data discovery, download, visualization, and analysis. *Environmental Modelling & Software, 37*, 146–156. doi:10.1016/j.envsoft.2012.03.013.

Bader, M. D., Ailshire, J. A., Morenoff, J. D., & House, J. S. (2010). Measurement of the local food environment: A comparison of existing data sources. *American Journal of Epidemiology, 171*(5), 609–617. doi:10.1093/aje/kwp419.

Bailey, R. G., U.S. Fish and Wildlife Service, & United States (1980). *Description of the ecoregions of the United States*. Washington, DC: U.S. Dept. of Agriculture, Forest Service.

Belobaba, P. P. (1987). Survey paper-airline yield management an overview of seat inventory control. *Transportation Science, 21*(2), 63–73. doi:10.1287/trsc.21.2.63.

Bishop, B. W. & Hank, C. F. (2016). Data curation profiling of biocollections. *American Society for Information Science and Technology Annual Meeting*, Copenhagen, Denmark.

Bishop, B. W., Grubesic, T. H., & Prasertong, S. (2013). Digital curation and the GeoWeb: An emerging role for geographic information librarians. *Journal of Map & Geography Libraries, 9*(3), 296–312. doi:10.1080/15420353.2013.817367.

Boone, J. E., Gordon-Larsen, P., Stewart, J. D., & Popkin, B. M. (2008). Validation of a GIS facilities database: Quantification and implications of error. *Annals of Epidemiology, 18*(5), 371–377. doi:10.1016/j.annepidem.2007.11.008.

de Bruin, S. D., Bregt, A., & Ven, M. V. D. (2001). Assessing fitness for use: The expected value of spatial data sets. *International Journal of Geographical Information Science, 15*(5), 457–471. doi:10.1080/13658810110053116.

Campbell, J. P., Gridley, T. S., & Muelleman, R. L. (1997). Measuring response intervals in a system with a 911 primary and an emergency medical services secondary public safety answering point. *Annals of Emergency Medicine, 29*(4), 492–496. doi:10.1016/S0196-0644(97)70222-1.

Cayo, M. R., & Talbot, T. O. (2003). Positional error in automated geocoding of residential addresses. *International Journal of Health Geographics, 2*(1), 1–12. doi:10.1186/1476-072x-2-10.

Chrisman, N. R. (1984). Part 2: Issues and problems relating to cartographic data use, exchange and transfer: The role of quality information in the long-term functioning of a geographic information system. *Cartographica: The International Journal for Geographic Information and Geovisualization, 21*(2–3), 79–88. doi:10.3138/7146-4332-6J78-0671.

Devillers, R., Bédard, Y., & Jeansoulin, R. (2005). Multidimensional management of geospatial data quality information for its dynamic use within GIS. *Photogrammetric Engineering & Remote Sensing, 71*(2), 205–215. doi:10.14358/PERS.71.2.205

Devillers, R., Bédard, Y., Jeansoulin, R., & Moulin, B. (2007). Towards spatial data quality information analysis tools for experts assessing the fitness for use of spatial data. *International Journal of Geographical Information Science, 21*(3), 261–282. doi:10.1080/13658810600911879.

Devillers, R., Stein, A., Bédard, Y., Chrisman, N., Fisher, P., & Shi, W. (2010). Thirty years of research on spatial data quality: Achievements, failures, and opportunities. *Transactions in GIS, 14*(4), 387–400. doi:10.1111/j.1467-9671.2010.01212.x.

Elliott, A. (2014). Top 10 Broadband Carriers Compete: DSL Kicking Tail. *IDinsight*. Retrieved from http://idinsight.com/featured/top-10-broadband-carriers-compete-dsl-kicking-tail/

ESRI. (2015). *U.S. Only Datasets*. Retrieved from http://www.esri.com/data/esri_data/what-you-get

ESRI. (2016). *ESRI Business Analyst*. Retrieved from http://www.esri.com/software/businessanalyst

Fayyad, U. M., Piatetsky-Shapiro, G., Smyth, P., & Uthurusamy, R. (1996). *Advances in knowledge discovery and data mining*. Cambridge, MA: MIT Press.

FGDC [Federal Geographic Data Committee]. 1994. National Geospatial Data Asset Management Plan. URL: https://www.fgdc.gov/policyandplanning/a-16/ngda-management-plan.

Goldberg, D. W. (2011). Advances in geocoding research and practice. *Transactions in GIS, 15*(6), 727–733. doi:10.1111/j.1467-9671.2011.01298.x.

Google. (2016). *Google Maps/Google Earth APIs Terms of Service*. Retrieved from https://developers.google.com/maps/terms

Graves, W., & Gerney, B. (2016). Notes on the inconsistency of subcounty scale demographic data: A comparison of five data vendors. *Papers in Applied Geography, 2*(1), 121–128. doi:10.1080/23754931.2015.1086413.

Grubesic, T. H. (2012a). The US National Broadband Map: Data limitations and implications. *Telecommunications Policy, 36*(2), 113–126.

Grubesic, T. H. (2012b). The wireless abyss: Deconstructing the US national broadband map. *Government Information Quarterly, 29*(4), 532–542.

Grubesic, T. H., & Mack, E. A. (2015). *Broadband telecommunications and regional development*. New York: Routledge.

Grubesic, T. H., & Murray, A. T. (2005). Geographies of imperfection in telecommunication analysis. *Telecommunications Policy, 29*(1), 69–94. doi:10.1016/j.telpol.2004.08.001.

Grubesic, T. H., Matisziw, T. C., & Murray, A. T. (2011). Market coverage and service quality in digital subscriber lines infrastructure planning. *International Regional Science Review, 34*(3), 368–390. doi:10.1177/0160017610386479.

Grubesic, T. H., Wei, R., & Murray, A. T. (2014). Spatial clustering overview and comparison: Accuracy, sensitivity, and computational expense. *Annals of the Association of American Geographers, 104*(6), 1134–1156.

Guptill, S. C., Morrison, J. L., & International Cartographic Association (1995). *Elements of spatial data quality*. Oxford, England: Elsevier Science.

Heipke, C. (2010). Crowdsourcing geospatial data. *ISPRS Journal of Photogrammetry and Remote Sensing, 65*(6), 550–557. doi:10.1016/j.isprsjprs.2010.06.005.

Hoehner, C. M., & Schootman, M. (2010). Concordance of commercial data sources for neighborhood-effects studies. *Journal of Urban Health, 87*(4), 713–725. doi:10.1007/s11524-010-9458-0.

InfoUSA. (2016). *Targeted Business Data*. Retrieved from https://www.infousa.com/product/business-lists/

Janssen, M., Charalabidis, Y., & Zuiderwijk, A. (2012). Benefits, adoption barriers and myths of open data and open government. *Information Systems Management, 29*(4), 258–268. doi:10.1080/10580530.2012.716740.

Jayakar, K., & Park, E. A. (2013). Broadband availability and employment: An analysis of county-level data from the National Broadband Map. *Journal of Information Policy, 3*, 181–200. doi:10.5325/jinfopoli.3.2013.0181.

Kashyap, V., & Sheth, A. (1997). Semantic heterogeneity in global information systems. In M. Papazoglou & G. Schlageter (Eds.), *Cooperative information systems: Current trends & directions* (pp. 1–24). Sand Diego, CA: Academic.

Kaufman, T. K., Sheehan, D. M., Rundle, A., Neckerman, K. M., Bader, M. D., Jack, D., & Lovasi, G. S. (2015). Measuring health-relevant businesses over 21 years: Refining the National Establishment Time-Series (NETS), a dynamic longitudinal data set. *BMC Research Notes, 8*(1), 507. doi:10.1186/s13104-015-1482-4.

Kerski, J. J., & Clark, J. (2012). *The GIS guide to public domain data*. Redlands, CA: ESRI Press.

Kitchin, R. (2014). *The data revolution: Big data, open data, data infrastructures and their consequences*. London: Sage.

Klein, G. (1999). *Sources of power—How people make decisions*. Cambridge, MA: MIT Press.

Koontz, C. M., & Jue, D. K. (2006). Public library facility closure: How research can better facilitate proactive management. *Public Library Quarterly, 25*(1–2), 43–56. doi:10.1300/J118v25n01_04.

Kruger, L. G. (2009). Broadband infrastructure programs in the American Recovery and Reinvestment Act (CRS Report No. R40436). Washington, DC: Congressional Research Service. Retrieved from https://www.acuta.org/acuta/legreg/l57.pdf

LaRose, R., Bauer, J. M., DeMaagd, K., Chew, H. E., Ma, W., & Jung, Y. (2014). Public broadband investment priorities in the United States: An analysis of the broadband technology opportunities program. *Government Information Quarterly, 31*(1), 53–64. doi:10.1016/j.giq.2012.11.004.

Lewis, K., Kaufman, J., Gonzalez, M., Wimmer, A., & Christakis, N. (2008). Tastes, ties, and time: A new social network dataset using Facebook.com. *Social Networks, 30*(4), 330–342. doi:10.1016/j.socnet.2008.07.002.

Madnick, S. (1999). Metadata Jones and the tower of Babel: The challenge of large-scale semantic heterogeneity. In *Proceedings of the 1999 IEEE Meta-Data Conference, 1*(13).

Miller, H. J., & Han, J. (2009). *Geographic data mining and knowledge discovery*. Boca Raton, FL: CRC.

Morrison, J. L. (1995). Spatial data quality. In S. C. Guptill & J. L. Morrison (Eds.), *Elements of spatial data quality* (pp. 1–12). New York: Elsevier Science.

Murray, A. T., Grubesic, T. H., Wei, R., & Mack, E. A. (2011). A hybrid geocoding methodology for spatio-temporaldata. *TransactionsinGIS, 15*(6),795–809.doi:10.1111/j.1467-9671.2011.01289.x.

Omernik, J. M. (1987). Ecoregions of the conterminous United States. *Annals of the Association of American geographers, 77*(1), 118–125. doi:10.1111/j.1467-8306.1987.tb00149.x.

Sherali, H. D., Lee, Y., & Park, T. (2000). New modeling approaches for the design of local access transport area networks. *European Journal of Operational Research, 127*(1), 94–108. doi:10.1016/S0377-2217(99)00325-2.

Sui, D., Elwood, S., & Goodchild, M. (2013). *Crowdsourcing geographic knowledge: Volunteered geographic information (VGI) in theory and practice*. New York: Springer.

U. S. Office of Management and Budget. (2010). *Coordination of geographic information and related spatial data activities*. Retrieved from https://www.whitehouse.gov/sites/default/files/omb/memoranda/2011/m11-03.pdf

U.S. Census Bureau. (2015). *TIGER/Line Shapefiles and Tiger/Line Files*. Retrieved from https://www.census.gov/geo/maps-data/data/tiger-line.html

Veregin, H. (1999). Data quality parameters. In P. A. Longley, M. F. Goodchild, D. J. Maguire, & D. W. Rhind. (Eds.), *Geographical information systems—Volume I, Principles and technical issues*. (2nd ed., pp. 177–189). New York: Wiley.

Walker, B. (2015). Everyday dig data statistics—2.5 quintillion bytes of data created daily. *V & C*. Retrieved from http://tinyurl.com/zhl9het

Walls, D. W. (2007). National Establishment Time-Series (Nets) Database: Coffee Analysis. *SSRN 1022333*. doi:10.2139/ssrn.1022333

Wang, X., Yu, C., & Wei, Y. (2012). Social media peer communication and impacts on purchase intentions: A consumer socialization framework. *Journal of Interactive Marketing, 26*(4), 198–208. doi:10.1016/j.intmar.2011.11.004.

Whitacre, B., Strover, S., & Gallardo, R. (2015). How much does broadband infrastructure matter? Decomposing the metro–non-metro adoption gap with the help of the National Broadband Map. *Government Information Quarterly, 32*(3), 261–269. doi:10.1016/j.giq.2015.03.002.

Zhu, X., & Davidson, I. (2007). *Knowledge discovery and data mining: Challenges and realities*. New York: IGI Global.

Chapter 8
Meeting Information Needs

Abstract Geographic information (GI) use spans the mundane activities of everyday life to the grandest scientific endeavors. Developing an increased understanding of GI access and use is possible using approaches that are grounded in human information seeking behavior. Furthermore, making GI more accessible, discoverable, and usable to various stakeholders requires an assessment of users' information needs and crafting services to facilitate GI access and use. This chapter will provide some background, terminology, and theory of human information seeking behavior. Instrumental to GI use are geographic information systems (GIS). The chapter concludes with a discussion of GIS software options and site license management approaches.

8.1 Original, Often-Ambiguous Queries

Most people, regardless of their background, culture or location, ask questions when an information need arises. "An *information need* is a recognition that your knowledge is inadequate to satisfy a goal that you have" (Case, 2012, p. 5). In most instances, a user attempts to address these needs by resolving them with incidental information, locating information within one's own knowledge, asking any nearby person (e.g., friend), consulting informal and/or formal resources (e.g., Google), or does not address the need with information avoidance behaviors. To satisfy some information needs, including many geographic information (GI) needs, a user may seek help from an information professional. Crafting questions about the unknown presents challenges because users do not know what they do not know, nor do they have the terminology to articulate the unknown. Some users assume GI exists to meet their information need in the exact way they imagine needing it, but this may not always be the case. The purpose of this chapter is to present researchers an approach to systematically study information needs, uses, and the software choices of users in the context of GI access and use.

Geographic information to satisfy many information needs *does* exist. Governments and other entities have invested in the creation of GI to answer common questions about the world around us. For example, how much oil is contained within a particular reservoir? How often is a particular land parcel underwater? What is the parcel worth? Who lives on that parcel? For some questions, users may

W. Bishop, T.H. Grubesic, *Geographic Information*, Springer Geography,
DOI 10.1007/978-3-319-22789-4_8

not have the necessary background information to negotiate their information need on their own. This is especially true for imposed queries (e.g., questions from an employer or teacher that were not self-generated) (Gross, 2002). Regardless of the information need, when asked, the task of any information professional in satisfying information needs is to move from a user's original, often-ambiguous to query, to providing an information match from an available resource (Taylor, 1968).

Human information seeking behavior research encompasses a wide array of interactions between users and resources. For example, two of the largest bodies of literature on information seeking behavior pertain to market research and political communication. Where the former is concerned, there is a focus on how to influence the purchasing decisions of potential customers and the latter seeks to address questions about how the media informs voting (Case, 2012). There are geographic aspects to such behavior, but the users (e.g., customers) do not necessarily seek GI and therefore do not fall within the context of GI seeking behavior. In addition, the majority of the work on human information seeking behavior is not concerned with everyday tasks. It primarily focuses on how users of specific information agencies seek information from their own resources (e.g., finding a relevant journal article in a library).

One of the major reasons that human information seeking behavior is of interest to this book is its links to reference work. Reference work began in libraries in the late nineteenth and early twentieth centuries as the result of several factors, including a surge in the volume and range of resources, an increased effort to use those resources, and the belief that many users needed professional assistance given the rise in the amount and the difficulty of using resources (Janes, 2003). A similar situation exists now with the emergence of the Geoweb and the associated surge in number and variety of GI resources. However, the relative ease of information retrieval, via the Internet or allied networked systems, means that almost any query produces search results. For users that are satisfied with any answer, an information professional is not needed. For many users, however, the value of information professionals' skills to vet the quality and authority of resources, as well as construct more precise and efficient search strategies, are appreciated because these skills produce more relevant findings and save users time. GI needs that lead users to ask for help from an information professional likely stem from complex questions that cannot be found without these skills. Furthermore, information professionals helping users may benefit from a thorough understanding of the way users seek GI and the typical motivations behind GI use. This intimate understanding also helps with search efficiency.

In his 1952 dissertation, Arthur H. Robinson called for a repositioning of cartographic study on the functionality of map symbolization and design. In part, this was because the ability to create GI "far outstripped our ability to present it" (p. 4). Today, the ability to create and present GI far outstrips the ability to organize it and study its access and use. GI access and use research should not be confused with the substantial amount of map use research done in Cartography. The factors for map use include "the perceptual and spatial abilities of readers, understanding of the symbol system (e.g., training or ability to understand the legend), goals, attitudes, viewing time, intelligence, prior knowledge, and preconceptions" (MacEachren, 1979, p. 5). Some of those same map use factors impact the ability of users to *find* cartographic resources, but the distinction must be made that map symbolization and design in Cartography are markedly different topics, with separate research questions and methodologies, when compared to the study of GI seeking behavior.

8.2 GI Seeking Behavior

GI is observable, ever-present, and relates to most everyday human activities. All of the geo-enabled information from location-based services (e.g., wayfinding and social media) and participatory personal information (e.g. activity trackers) present researchers ample avenues for investigation. However, this discussion will *not* focus on the needs and seeking behavior from these avenues because the geographic aspects of those information types are secondary and in most cases inconsequential to its use. Instead this discussion of GI seeking behavior will center on a user's search for any object (i.e., document) that contains geographic representations. For example, this will include things such as analog (i.e. print) cartographic resources, digital geospatial data, or any other information object that captures a place in space-time with a categorical theme. Buckland (1991) uses cartographic resources as a series of examples in support of his definition of information-as-thing. "If a map is a document, why should not a three-dimensional contour map also be a document? Why should not a globe also be considered a document since it is, after all, a physical description of something?" (p. 354). Goodchild (2003) asserts that GI is an *atomic element* composed of two qualifiers, a physical location in space-time (i.e., latitude, longitude, altitude, and time stamp of occurrence) and another attribute related to that physical location. This GI definition and earlier theories from Geography, such as the *geographic matrix* (i.e., geography characterized by location, time, and attributes) matches with Information Science's physical paradigm definition of information-as-thing where GI is viewed as a tuple of a location (x, y) and a property (z) (Berry, 1964).

For this discussion on GI seeking behavior, both quantitative and qualitative GI should be conceptualized as inherently physical information objects. In addition, some of the same needs and behaviors may exist in the search and use of astronomical data, but the following information seeking behavior scenarios will be restricted to GI below the mesopause as an atomic element. Goodchild does not demarcate a minimum scale for GI and states that the atom size is contingent on "the number of distinct properties at any location" (Goodchild, 2003, p. 23). This conceptualization includes geographic representations at relatively larger ratios than objective reality (e.g., magnified fire ant mound architecture), but examples of that GI type are rare and most users will not have subhuman GI needs (Cassill, Tschinkel, & Vinson, 2002).

GI in any of its physical forms is central to the decision-making of governments, industry, organizations, science, other entities, and individuals. A more productive discussion of GI seeking behavior requires the context of specific scenarios to highlight a user's GI need and their actions to fulfill that need. Many information agencies across sectors expedite GI use, but the setting of an academic library presents a well-defined user group (i.e., faculty and students) that is staffed with dual-purpose information providers that simultaneously help users locate GI and also teach users how to find and use GI. Furthermore, situating these GI seeking behavior scenarios within an academic library has merit. A recent survey of faculty and graduate students at a land-grant university indicated that 50.9% use geographic information systems (GIS) (131 of 257 respondents across academic departments), and it is not likely that that level of GI user saturation exists in other communities nor would it include such a wide range of users' skill levels (March, 2011).

Within this academic setting, consider how information professionals and related services respond to user information needs. The response generally falls into two categories, *reference* and *instruction*: (1) answering user questions by locating GI; (2) instructing users how to find, evaluate, and use GI; and (3) creating tools to aid the search for GI without the assistance of an information professional. In the following subsections we explore both categories, reference and instruction, in detail.

8.2.1 Reference Scenario

Bishop (1915) presented a paper at the American Library Association (ALA) meeting to define reference work for the first time as "the service rendered by a librarian in aid of some sort of study" (p. 134). Librarians have long equated the question-negotiation process with "detective work," teasing out an information need from a user's question (Walford, 1978, p. 89). In order to provide a more formal presentation of reference skills, the Reference and User Services Association (RUSA) published guidelines for Behavioral Performance of Reference and Information Service Providers to use in the training, development, and evaluation of reference services (American Library Association and Reference and Adults Services Division, 1996; Reference and User Services Association, 2004). The five sections of the RUSA guidelines—approachability, interest, listening/inquiring, searching, and follow-up—provide information providers with common sense recommendations for satisfying user needs.

Unfortunately for librarians, the "detective work" and expert knowledge of information resources required to answer complex and engaging research questions is not always needed. A recent study of chat reference (i.e., text-based messaging in real-time with an information professional) found 30% of questions were ready reference (i.e., questions with simple factual answers that could be found through encyclopedias, almanacs, or through search engines and databases) (Connaway & Radford, 2011). In addition, another study found 11.5% of the total questions asked at all reference service points on one university's campus were simple wayfinding questions (e.g., where's the bathroom?) (Bishop, 2012). Still, GI reference questions that require the expertise of information professionals do happen with some regularity. Librarians also spend time troubleshooting GIS software issues with users not related to locating GI and focus on teaching users what to do with GI once found.

Nearly all reference questions asked to information professionals working with cartographic resources relate to a place. For example, when a librarian finds a reference map, of a place, for a user, this can help address information needs related to locating and identifying important real-world features. Most location-based services and Geoweb maps provide a current snapshot, but some users may be seeking historical information. As a result, alternative digital and cartographic resources may be more appropriate. Not surprisingly, one study found that 70% of GI requested at one academic library were about the state where that library was located (Scarletto, 2011). As a result, it seems that geographic proximity likely biases information needs across many information agencies (Scarletto, 2011).

Unlike other reference interactions, less ambiguity exists for users seeking GI as they typically know what they would like to find. For example, few GI inquiries are so unique or novel that the resource will not exist. Other users or entities (e.g., scientific teams) likely have similar interests, which result in the existence of a geospatial data collection that represents a particular geography. In some instances the *exact* GI or map may not exist because cartographic design constraints limit the number of variables that can be presented legibly on a map. Beyond these cartographic limitations, there are several other reasons that a particular document or dataset may not exist. First, some GI may not be collected because the technology (of the time) would not permit it. For example, there is no aerial photogrammetry of the U.S. Civil War. Second, there may be differences in the GI that the user wants (e.g., time, place, and scale) versus the GI that is available (e.g., time and place are ok, scale is not). These mismatches between need and availability may impact fitness for use. Finally, some librarians suggest that users are disappointed when a map of their GI need already exists because the joy of mapmaking and leaning GIS skills outweigh the time saved by meeting their needs with an existing cartographic resource. That said, some GI needs will always require primary sources because the original maps trump any newly created substitute, especially for historic GI.

Consider a simple scenario. A student in the discipline of Information Science has a class assignment asking him/her to find the oldest maps of the Reading Terminal in Philadelphia *available online*. In this instance, it is clear that a host of clarifying questions need to be addressed to best match this need with a specific cartographic resource. Imposed queries are common in academic settings as instructors incentivize students to seek out particular information to gain new knowledge through various assignments. For the academic librarians at this institution, answering these queries over and over again creates the appearance of clairvoyance. In comes a confused student, casually approaching the reference desk—and there are only a few potential reasons leading them to ask for help. Obviously, an information professional does not need psychic powers to discern that a student needs assistance. But, there is a process through which the information professional clarifies which sources the user has already consulted in an effort to guide the student to the next logical resource.

For a user, the imposed query for an assignment is an information gap that changes throughout the search process, especially as new information sources are consulted. Case (2012) delineates the three common factors in any information seeking behavior—(1) sources of information; (2) time; and (3) degree of thoroughness. These factors will help analyze the Reading Terminal information seeking behavior. The degree of thoroughness is less important for any imposed GI query, but especially when it comes to the Reading Terminal inquiry. There will be no loss of lives and/or financial ruin from the use of inaccurate GI in a class assignment. For this particular scenario, the thoroughness of the answer to this question is not a huge concern because the assignment comprises many other questions—most of which the student believes are answered correctly. Of course, time pressure impacts how quickly an information need must be addressed. For this assignment, time is the most pressing factor because the given assignment is due the following morning. Luckily, the source of information to consult is known to the student because it was covered in course lectures and readings. The commonly used resource for historic GI—Sanborn maps—should provide an answer.

For most searches, students (and most users, generally speaking) begin with a Web search engine, even though the resource that could provide the answer is already known and a less circuitous route exists for finding the required GI. Through a combination of web crawling, indexing, and page ranking, Web search engines provide information retrieval services to an ever-expanding array of open Web pages, images, and other file types (Morville & Callender, 2010). For much of the Geoweb, and in particular authoritative GI resources, the functionality behind most search boxes do not yet have an algorithm to adequately address the dual complexity of an attribute symbolized in text and geographic contexts for GI searches. Rather, a search engine can pull up the Web pages for GI sources whereby a user can subsequently search portals, warehouses, and repositories to meet their tangible GI need.

When the student uses Google to find information on Sanborn maps, the first result is ProQuest's *Digital Sanborn Maps, 1867–1970* (http://sanborn.umi.com/); however, these proprietary resources require a log-in. The student then clicks on another page from the search engine results and visits the Library of Congress (http://www.loc.gov/rr/geogmap/sanborn/), but most of the Philadelphia maps are not available online. At this point, the student's Information Science training kicks in and he/she attempts the search again using local library digital resources, but since Philadelphia is not nearby, the university library only purchases access to the local states' Sanborn maps. No maps from Pennsylvania are available. This generates a shift in the student's information seeking behavior, especially with the due date approaching, and he/she reverts back to the bad habits employed before information professional training. The student turns to Wikipedia with the hope of finding a link to resources in Pennsylvania (http://en.wikipedia.org/wiki/Sanborn_Maps). Unexpectedly, this works because the Wikipedia page links back to Penn State University's Libraries' website. This site includes the needed Sanborn maps (http://www.libraries.psu.edu/psul/digital/sanborn.html). From the Penn State resource, the student uses a chat reference service to ask a librarian at that university how to use their resource. The librarian via chat helps the user locate a pdf of the index page from the oldest Philadelphia Sanborn maps available—1916—to search for the number map sheet that has the indexed Reading Terminal.

With either index page approach, the information seeking behavior does not stop with locating the proprietary, digitized Sanborn maps resource even though that part of the process presented a number of challenges for the student. The GI need is not met until the student navigates the archaic index of the original print resource, which is frustrating for digital natives accustomed to the indexing granularity and zooming and panning functionality of the Geoweb. This is especially true at the end of a long search for the appropriate and available GI resource. In scenarios where a user will not have the imposed query as a motivator, clunky tools without keyword search functions, even though online, may impede GI access and use and prevent others from taking the proverbial ride on the Reading. In addition, most users do not take GI data discovery coursework and the behavior of those users would likely include asking an information professional for help.

8.2.2 Instruction Scenario

Distinguishing between reference and instruction in an academic setting can be challenging. For example, teaching users to seek GI on their own must occur as part of any question-negotiation process. In 1999, the Association of Research Libraries (ARL) Geographic Information Systems Literacy Project surveyed 121 ARL member institutions. The results of this survey suggested that half of all GIS users needed extensive assistance (Davie, Fox, & Preece, 1999). Where extensive assistance stops and instruction begins is difficult to assess.

There are many types of instructional activities occurring in all information agencies. For example, throughout libraries (e.g., school, academic, public, and so forth), bibliographic instruction has transitioned toward *information literacy* and relates to teaching users how to find, assess, and use information. The Association of College and Research Libraries (ACRL) (1989) *Presidential Committee on Information Literacy: Final Report* defined an information literate person as one that had the ability to locate, evaluate, and use effectively the needed information. These are critical skills to succeed in a world with a seemingly infinite number of resources. GI academic librarians offer a wide variety of instructional services, such as GIS training, community exhibits, and GIS Day celebrations, and these services often include outreach to the K-12 schools and broader communities (Weimer, Olivares, & Bedenbaugh, 2012). For example, librarians at one institution have provided aerial photographs to K-12 students to teach about changes in land use over time as well as inform them of the value of preserving this GI (McAuliffe, 2013). Another group of librarians provided students with a series of orienteering workshops using the libraries' large scale topographic USGS maps and university global positioning system (GPS) equipment and GIS software (March & Darnell, 2012). The future of GI instruction for information professionals is limitless given the growth and reduced costs of new technologies (e.g., drones) and data (e.g., VGI).

In this next scenario, an undergraduate Geography student needs help completing a GIS lab. This scenario is typical for information professionals working in academic libraries. The information seeking behavior of the undergraduate student who is having trouble with a GIS lab on geocoding should be relatable for many readers and hopefully does not result in any emotional flashbacks. This undergraduate student seeks instruction assistance in an academic library for a number of reasons. Again, it is useful to use Case's (2012) common factors that influence information seeking behavior. The sources of information consulted for completing the lab include the course lab book, Web pages from the publisher, fellow classmates, the instructor, and an information professional. The GIS lab is an imposed task from the instructor and due in only a few hours. Therefore, there is little time for the student to learn how to accomplish the core tasks in the lab. Further, the degree of thoroughness is less of a concern for this student because the labs have been generously graded pass/fail. In this instance, the student simply needs some points to maintain his/her grade in the course.

At first, the student follows the course lab book instructions, step-by-step, to geocode addresses. For whatever reason, the address locator recommended in the book for geocoding tasks is unavailable. Worse, the address locator that is available keeps crashing. Frustrated, the student consults support pages from the publisher's Web portal and learns that the geocoding tasks require a special online account to access that software functionality. This was confusing because until the geocoding lab, a public online account provided the tools necessary to complete all of the other labs for the course. To seek the instructional information needed to complete the lab, the student posts a question on the class discussion board because there is no class meeting prior to the assignment deadline. Although the instructor provides other online geocoding options to complete the lab, there is not enough time to create the special online account for the class to access the software. Another student suggests using the CD that came with the book and its address locator, but the student does not have an optical drive.

For whatever, reason, the student does not pursue either workable option, instead insisting on learning how to geocode exactly as the book instructs. The student goes to the library to ask an information professional. The staff member asks about this particular instance of information seeking behavior and determines a course of action. Fortunately, the staff member has the ability to set up the special online account for the student. With access to the online address locator, the student completes the lab exactly as the course lab book instructs and turns in the lab on time. Of course, there was also some hands-on help from the staff member. In this scenario, as with many real-life instruction information seeking behaviors, the teaching relates to showing users how to use the information technology and not just the information itself. Instruction by information professionals also includes teaching users how to locate, evaluate, and use GI, but the key takeaway from this section is to see the GI seeking behavior from the user's perspective. In most instances, a user will have consulted other resources before asking an information professional for help, has a time constraint of some kind, and, especially in academic settings, has some flexibility in their degree of thoroughness. This is a key difference, because within information agencies beyond academic settings, the degree of thoroughness overshadows the time constraint (e.g., building construction). For another example, a student may need peer-reviewed articles for a final paper and an information professional can help them with that information need as well as take a moment to give the student a mini-lesson on scholarly literature. Reference and instruction are both active roles for information professionals, but the final scenario presented is a more passive, yet valuable, service offered in most academic libraries.

8.2.3 *Information Guide Creation Scenario*

In the previous two scenarios that detailed GI seeking behavior, both scenarios generated interactions between users and information professionals. In an effort to empower users to answer their own GI questions and reduce the questions requiring real-time expert assistance, information professionals have a longstanding practice of creating guides to resources. In the past, users of collections physically entered information agencies and consulted handouts and bibliographies. Today, access to online pathfinders help users traverse the quagmire of available online sources and find the highest quality GI without leaving their home. A popular tool for building guides in academic libraries is Springshare, Inc.'s LibGuides because of its tab-based structure, built-in Web 2.0 features, and ease of use for users and builders of the guides (Gonzalez & Westbrock, 2010). A concern shared by many information agencies employing these guides relates to their management, ensuring they are accessible by users with varying literacies and technical capabilities, and keeping the GI current (Smith, 2008). As the following scenario will highlight, the temporal dimension of GI presents challenges for GI seeking behavior. Using Case's (2012) common factors, the sources consulted in this scenario are numerous, the time pressure irrelevant, but the degree of thoroughness important.

Consider, for example, a local historian that decides to create an online "life map" representing all the known whereabouts of his/her state's first governor. Given shifting political boundaries over time and place-name changes, these types of information seeking behaviors present an extremely complicated suite of needs because the availability of historic documents varies widely. In addition, primary resources related to important political operators business people or other historical figures are sometimes missing because of document destruction (e.g., courthouse fire), improper preservation practices (e.g., acidic paper), or no original documentation at all. These searches are also more problematic in regions with conflict and unrest that cause place-names and political areas to change. Luckily, the local historian found a letter in a local museum's holdings that will help create the life map. The Commissioner of the Railroad and Public Utilities of the State of Tennessee compiled the seven places John Sevier lived in response to a 1941 request from an interested descendant.

The Commissioner's letter indicates Sevier was born in Sullivan County, Tennessee. However, the local historian, having read multiple biographies, knows Sevier was born in Virginia. In fact, in a book indexing Virginian historic markers, he/she finds two historic markers in Rockingham and Shenandoah Counties claiming to be Sevier's birthplace. This presents a unique GI problem as he/she does not understand how a person could be born in two places. After consulting Wikipedia, the mystery was solved because both Counties were once apart of Augusta County, Virginia. Over time, Augusta County has been divided into at least ten current Virginia counties, as well as most of West Virginia and Kentucky. Still, the town of New Market, where Sevier was born, never moved. It still exists within present-day Shenandoah County and can be easily georeferenced for the life map. The early life of Sevier also included a stop in Fredericksburg, Virginia, which remains in the same location. Without references, the Commissioner's letter has lost all credibility and is a good reminder to the local historian to keep a critical eye on secondary sources. In defense of the Commissioner, the information in the letter was compiled through correspondence with a judge and he did not have access to any search engines in the 1940s. The letter does place the Seviers in Tennessee in 1773 and those locations are verifiably accurate through several of their deeds.

The next three residences in the letter are in present day Sullivan, Carter, and Washington Counties, Tennessee; however, in Sevier's time the geopolitical history was complicated by deeds using the metes and bounds surveying methods and the frequent subdividing of counties. For example, the local historian found one Sevier deed from the State of North Carolina that outlines a property as follows "a Sweet Gum Tree, a Poplar Tree at the top of Lynn Mountain, or a clump of Ashes by the road along the river". Therefore, only approximate point locations of Sevier residences near remaining rivers can be added to the life map. The local historian, knowing early Tennessee history, understands that the first non-native settlers purchased lands from the Cherokee to form the Watauga Association. Then, the British admonished the legality of those transactions, along with a few additional twists and turns, that included battles with both North American indigenous peoples and European colonists. These clashes and the outcomes made Sevier famous and he parlayed battlefield success into a career carving out one proposed state (Franklin) and later the creation of a territory that did become the 16th state of the U.S. (Tennessee). Documenting the locations of all those travels would add value to the life map, but Sevier's tumultuous life and little detailed documentation allows for it.

In dismissing the Commissioner's letter for several inaccuracies, the historian turns to a thesis on Sevier to complete the remaining points on the life map. The thesis outlines Sevier's life starting in 1796, when he served as governor. Since all of Sevier's residences occurred in one area, greater accuracy is possible (Barber, 2002). In addition, the original capital of Tennessee—Knoxville—has preserved many of the same streets in its Old City allowing for modern geocoding of the remaining Sevier homes. First, Sevier rented a house at the corner of Cumberland and Central Avenues and although that rental unit is long gone, the city block remains is the same. He later moved to Marble Springs, a home that is still preserved in the County that bears his name. Marble Springs is on the U.S. National Register of Historic Places with Sevier's cabin still maintained in its original location. This last location completes the local historian's life map of residences, but does not include final resting places.

The local historian dug further into a recent biography to learn Sevier died after attending a feast at the Green Corn Dance while working to survey land in western Georgia that a few years later became Alabama. For 74 years his remains were buried near the Tallapoosa River until he was re-interred at the County Courthouse in Knoxville, Tennessee in 1889 (Belt & Nichols-Belt, 2014). If the local historian so desired, the graves could easily be georeferenced as accurately as the residences. Sevier's eventful, seventy-one year long life presents the local historian several challenges in locating GI for all the places he lived and was buried, but highlights the challenges facing those working in the geohumanities. The haphazard GI seeking behavior required to locate all the dispersed historic GI sources for use would benefit from some streamlining.

By simply asking an academic librarian or finding the LibGuide, Historic Geographic Information Systems (HGIS) resource at the University of Tennessee-Knoxville, the local historian could have saved some effort in locating GI for the life map and GI resources that capture the changes to the political boundaries over time. The guide compiles relevant links to historical maps of the area, historical administrative boundaries, aggregate Census data, and other useful datasets for historical research in a centralized place. Hopefully, more of these types of guides can be created and maintained to help meet future GI needs, especially for the 16 National Geospatial Data Asset (NGDA) themes' communities (e.g., biodiversity and ecosystems, geology, and so forth). This work should reduce the complexity of other GI seeking behaviors.

8.2.4 The Need for a Theoretical Framework of GI Access and Use

Given the pragmatic demands of satisfying information needs adequately and quickly in information agencies, the theoretical frameworks of Information Science are not always structured to meet demand and inform practice in the real-world. Within academic libraries and at other information agencies, for example, experts who help a community of users with similar information needs may not see the utility of models and theories to study these everyday interactions. This is not surprising, given that theories and models are not necessarily structured for solving day-to-day problems in these environments. For example, the purpose of theory is to explain a phenomena based upon certain assumptions, principles, and relationships. The purpose of a model is to sequence proposed stages in order to describe, predict, and explain phenomena. Once validated, a model may spawn a theory, but with constant changes to information and communication technologies and the information available, information seeking behavior research does not have comparable theories to those developed in other social science disciplines. As a result, the academicians studying human information seeking behavior, in all contexts, have seen the need to create many models to lead to the systematic study of these phenomena. The Association of Information Science and Technology (ASIS&T) monograph, *Theories of Information Behavior*, provides readers an introduction to some of the many existing information seeking behavior models and theories (Fisher, Erdelez, & McKechnie, 2005).

Many of the models in Information Science reflect the flow and structure of a message from a sender, through a channel, to a receiver, with the potential for feedback (Shannon & Weaver, 1949). Consider, for example, Wilson's models of information behavior (1981, 1997). Wilson contextualizes the information needs by separating out factors related to a user's life world, the information systems employed to address an information need, and other resources consulted in a search. Additional categories added to the models over time help to theorize the complexities of human information seeking behavior (Wilson, 1981, 1997). Nearly all the models in Information Science are inherently or explicitly built upon the underlying philosophical assumptions of the Sense-Making metatheory. The Sense-Making metaphor of a human's cognitive process includes where a person has been (e.g., experiences), where the person is (e.g., their current information gap), and where a person is going (e.g., consequences) (Dervin, 1983). The philosophical assumptions underlying the Sense-Making model include (Dervin, Foreman, & Lauterbach, 2003, p. 270):

1. Both humans and reality are sometimes orderly and sometimes chaotic;
2. There is a human need to create meaning, and knowledge is something that always is sought in mediation and contest; and
3. There are human differences in experience and observation.

A theoretical framework of GI access and use should adopt these underlying assumptions from Information Science. In addition, a theoretical framework that conceptualizes information-as-thing should include two elements of Tobler's First Law of Geography: (1) everything is related to everything else, and (2) near things are more related than distant things (Tobler, 1970).

As detailed earlier in the book, all GI is composed of atomic elements, very often consisting of at least one location (x, y) and at least one property (z). For users with GI needs, sense-making occurs to fill the knowledge gap through the lens of their world view, a compilation of their past experiences, identities, thoughts, attitudes, emotions, memories, culture, domain knowledge systems, and present circumstances. To assist a user, an information provider must assess the location or locations, as well as any related attributes within a user's information need to understand their unique context and to provide appropriate, satisfactory, and relevant GI.

To research these phenomena, researchers have employed many methods such as descriptive analyses, obtrusive methods, unobtrusive methods, observation, conjoint analyses, and cost-benefit analyses. However, all of this work relies on the same theoretical framing. Namely, through the basic assumptions that underlie the actions occurring when a user with a knowledge gap addresses their information need with mediation from an information professional (Matthews, 2007). Because all humans use GI, issues of GI access and use permeate all the worlds' systems. However, human information seeking behavior for GI that does not consist of everyday wayfinding should receive more systematic study. For example, what GI are found (or not), how GI are found (or not), and when GI are found (or not) impacts all of us and are worth considering in a more formal manner.

8.3 Using GI

The geospatial tools available for anyone using GI once it is accessed continue to evolve with the influence of the Geoweb. However, many employers are still seeking job candidates that have an ability to operate proprietary GIS software packages. Many, in fact, list these abilities as a requirement in the GI job descriptions. GI organization, access, and use are also expected skills, but these specific skills may not be directly linked to specific software in job ads. Worse, employers may assume that candidates who can operate a proprietary GIS package have these skills, anyway. Complicating matters is the appearance of more free and open source (FOS) GIS packages (Ramsey, 2007). Also, the Geoweb changes user expectations and moves many proprietary software products toward web-based online platforms. These platforms have limited functionality compared to some desktop versions, at least for now. To provide some perspective on the use of GIS software platforms, and their role in GI use, we outline some basic details of several open source and proprietary GIS packages, as well as brief case study from the administration of one proprietary GIS software license to contextualize the importance of tool access for GI use. In short, the ability to gain tool access is an often overlooked precursor to GI use. GI use for many users goes beyond mobile location-based applications (apps) and Geoweb services. Further, although open source options may not always fulfill the needs of individuals who require robust software for specialized spatial analysis, open source tools are improving rapidly.

8.3.1 Free and Open Source

Open source GIS is not new. The U.S. Army Construction Engineering Research Laboratories created the Geographic Resources Analysis Support System (GRASS) in 1985. There are many options for open source GIS software (e.g., FreeGIS.org lists 356 at present). Software packages have different origin stories with some started by commercial companies (e.g., uDig), some started as research projects (e.g., GRASS), others created by GIS enthusiasts (e.g., Quantum GIS), and even more supported by governmental agencies (e.g., gvSIG).

FOS GIS allows anyone to legally download and experiment with GIS software without any licensing fees. For obvious reasons, fees can be a financial barrier for some users and organizations. In some information agencies (e.g., libraries) with tight budgets, this fiscal restraint has led many to switch many information technology operations to open source options (e.g., Online Public Access Catalog (OPAC)) (Payne & Singh, 2010). In developing countries, open source tools enhance the ability of governments to share GI with more residents and encourage e-government engagement (Williams, Marcello, & Klopp, 2013). Beyond the obvious financial benefits, other advantages of open source include its unrestricted use and its support of other open standards. The Open Source Geospatial Foundation (OSGeo), established in 2006, assists in the collaborative development of open source geospatial software and promotes its use (http://www.osgeo.org/). As one might imagine, OSGeo provides the necessary financial, organization, and legal infrastructure to help developers to *compete* with proprietary software options. With 279 charter members of OSGeo, the treasure trove of code available to its users provides a relatively large sandbox for GIS development. R, the open source programming language and software development platform for statistical computing serves as another example. R enables spatial analyses through a community of programmers sharing a number of computational tools (e.g., http://r-gis.net/). Scientists and researchers have plenty of options for FOS GIS. The open source approach enable those resourceful enough to build upon existing knowledge and create new processes, tools, and knowledge.

Still, license fees for proprietary software do impart several advantages that most GIS users find worth the extra cost. First, proprietary software does not require users to learn to code. Even when open source GIS is designed for "out of the box" use, some find the learning curve too steep. In one heuristic usability test of GRASS, QGIS, uDig, gvSIG, OpenJUMP, MapWindow, and ArcGIS, Donnelly (2010) found that when creating thematic maps, the expert user found proprietary software was easier for reprojecting data and automatically labeling features when compared to alternative, open source options.

One hidden cost of all tools is the time it takes for users to master GIS concepts and software platforms. In the heuristic usability test above, the expert user may unconsciously have performed better in ArcGIS software because of a less steep learning curve. Students from institutions with Esri site license courses may have a head start over students with access to only open source platforms, even if they include accompanying educational materials and associated illustrations, because employers expect this expertise. Obviously, not all of the skills gained from learning one GIS software platform are easily translatable to others, but if the learning process is supported with good instructional resources it can reduce the frustrations of new users and improve learning potential for other platforms. Similar to training costs, the deployment of software across multiple university departments, colleges, and learning hubs, along with troubleshooting any problems, can become costly. Although cloud computing removes some of the physical requirements and time involved in the distribution process, managing access tools for this software is a job that requires a dedicated information professional (e.g., systems analyst). Most users prefer spending their valuable time learning and using GIS, rather than the time-consuming tasks associated with obtaining and troubleshooting software.

8.3.2 *Proprietary Software*

For many that find coding cumbersome, or organizations that determine the cost-benefit of FOS does not favor that option, the choice of proprietary software makes more sense. It often includes a warranty, the components work seamlessly with other proprietary software, and the documentation is excellent (Steiniger & Bocher, 2009). Again, potential employees may notice that many organizations/employers list knowledge of specific, proprietary software in job descriptions. The software listed the most by U.S. employers are those from the Environmental Systems Research Institute (Esri). To date, when totaling all versions of Esri's ArcGIS for Desktop, more than one million users have installed the software in more than

350,000 organizations, including most U.S. federal agencies and national mapping agencies, 45 of the top 50 petroleum companies, all 50 U.S. state health departments, and most forestry companies (Esri, 2015). In 2010, the ARC Advisory Group estimated Esri's worldwide market share at 40.7% of all GIS software types that perform basic mapping functionality, to advanced spatial analysis (Reiser, 2010). To a certain extent, this level of market dominance by Esri forces most colleges and universities in the U.S., especially those that house several information hubs (e.g., academic/special libraries, archives, museums, and so forth) to purchase an Esri site license. Their belief is that accessibility to this software will help students obtain jobs in the public and private sectors, as well as facilitating faculty and student research.

Open source tools fit the needs of many GI users, but there are several reasons proprietary commercial products exhibit some level of "stickiness" in academic environments. In addition to providing superior customer service and quality educational materials, it is the perpetuation of educators teaching the tools they already know. This gives established commercial software an advantage over the multitude of FOS GIS options and the growing plethora of versatile Geoweb tools in the vast majority of U.S. higher education institutions.

For proprietary software, each institution manages a site license with an individual or team that takes responsibility as gatekeeper(s) to legally manage software access. The administrator deployment approach controls who benefits from GIS access and GIS use. In short, site licenses are attempts by vendors to institutionalize use terms as standard operating procedures for consumers. Vendors create use terms to define end-users' rights.

Interestingly, the management and distribution of scholarly communication at higher education institutions provides some parallels to GIS software administration. In many instances, a third party, like a campus administrator, ensures that an institution complies with a site license agreement for accessing and distributing the copyrighted journal material. However, the vendors and end-users have little influence or understanding of the local distribution and management approaches. For databases of journal articles and e-books in U.S. higher education, academic librarians serve almost exclusively as the intermediaries to proprietary content and act as intellectual property stewards. In addition to tasks already outlined, academic library staff must ensure secure remote access to content for affiliated users only. This work increases staffing costs beyond the content or software purchases. These services also are hidden value for end-users, but institutions that understand the importance of access to tools, data, and scholarship for their users' education invest to make desired and beneficial information as available as possible.

With proprietary software, the units charged for administration have not been as thoroughly studied as journal access in academic libraries. Thus, it is worth exploring how this type of software is managed in university settings and how this may (or may not) impact GI access and use. Specifically, we examine the use terms for the Esri education site license. With a unique use term for institution-wide access, Esri effectively removes the burden of seat-limits, a problem that many end-users and campus software administrators have grappled with in the past. Esri's site license program includes software and software updates, access to online courses, Esri User Conference passes, and technical support services. Despite this uniformity in the use terms, each institution administers access to this same software differently based on situational factors across campuses. In sum, the Esri education site license is a representative case study to get a snapshot of the dissemination of GIS software across higher education.

8.3.3 GIS Site License Administration

In 2015, upon institutional review board (IRB) approval, a recruitment email was sent to approximately 50% of Esri's U.S. higher education licenses campus administrators (randomly selected) to determine administrative practices (79 of 308 responded; 25.6% response rate). Participants completed a survey asking the following administrative and demographic questions:

(a) Which of the following Esri products do you distribute and manage?
(b) How does software distribution occur?
(c) Which of the following best describes your institution?
(d) What is the size of the student body at your institution?
(e) How many years have you been distributing software?
(f) How many years have you been working with GI?
(g) Please select the answer choice that indicates your credential and/or degree. (Select all that apply).
(h) Please provide any additional feedback you may have on managing and distributing software.

The following subsection details the findings on how software distribution occurs.

8.3.3.1 Findings on Software Administration

All responding participants deployed ArcGIS for Desktop, and 75% deployed ArcGIS for Server, ArcGIS Online (AGOL), and other products. User request was the most common selection for software distribution with 92.3% making the selection. Still, 78.9% chose department request. This shows distribution occurs in several ways. The next questions concerned the type and size of institutions. Doctorate-granting University represented 65.4% (334) of the institutions responding, with only 26 as Master's College or University, 18 as Baccalaureate College, and 14 identifying as Associate's College (note: Esri categorization of institution type). Nearly half of participants self-selected a size of very large (more than 20,000 students) (46.2%).

The demographic questions provided insight into the campus administrators. Most participants had been distributing the software for 6 to 10 years (32.7%) with those with more than 15 years (30.8%) a close second. The third most selected group were those who had been doing it for 11–15 years (25%). The remainder were shared between those working 3–5 (9.6%) and 0–2 (1.9%). Clearly, many participants had been doing this job throughout many iterations of GIS software and distribution. Most participants had been working with GI for more than 15 years (57.7%). For the final demographic question, the majority (53.9%) had a degree in a geo-related field (e.g., Geography, Geology, Earth Sciences, Urban Planning, and so forth). The other degrees varied widely from Computer Science to Engineering to Library Science and two indicated they did not have any degree.

Many of the findings followed the common assumptions made that larger institutions distribute more and to more departments than smaller ones. For example, for type of institution by product the Doctorate-granting Universities were more likely to deploy ArcGIS for Server, provide online training courses, and distribute more than 100 online training codes. The very large institutions (>20,000) were more likely to distribute AGOL and ArcGIS for Server and the very small institutions (<7000) are less likely to distribute those products.

Yet, the size of institution did not influence distribution approaches. The approaches varied across type of institution having some distribution differences. All Baccalaureate College participants selected only department request for all distribution. This approach effectively removes access to any end-user not affiliated with a department teaching GIS.

The years of experience had little impact on distribution approaches, but campus administrators who had been working with GI for more than 15 years were more likely to distribute the products of AGOL, ArcGIS for Server, and other site license items. In addition, campus administrators with more than 15 years of GI experience were more likely to distribute by department request. Qualitative methods are needed to determine why this distinction exists, but we speculate that experience leads to familiarity with more products and their subsequent promotion on campus.

Finally, the survey concluded with an open-ended question to gather distribution suggestions. For institutions that have now implemented a central server for distribution of software, the process is becoming streamlined with authorization files that both first-time and annual renewal users may access on their own. GIS software access is proliferating, but not actively promoted across campuses. However, some institutions do not support this central server option. Interestingly, some institutions still preferred distribution via DVDs, as some users' off-campus connectivity makes central server downloading problematic or impossible.

8.3.3.2 Considerations for Software Distribution Practices

For much of the GI workforce, the individuals and agencies who control access to GIS software determine who is trained to do GI jobs, who may do GI research, and who may use GI. The implications for students are significant, especially given projected growth of the geoservices industry. Certainly, the entity purchasing and providing access to any product on campus has different motivations than vendors or end users. In addition, the more widely a software platform is deployed, the more costly it is to manage, although economies of scope and scale can lower the overall distribution costs. Since the campus deployment models vary so widely, allowing academic libraries to serve as the gateways (and gatekeepers) to GIS software makes good sense. Librarians have a history of providing access to other campus-wide resources, including e-journals. In turn, academic libraries would likely find success in distributing GIS software in the most discipline agnostic manner possible. Specialized software other than this particular GIS tool may also be tightly controlled, but as knowledge work explores the interstitial spaces between disciplines, a more open mentality to tool access coupled with open data and open access journals should remove some obstacles to these endeavors.

With AGOL growing into the central Esri product, it is clear that software is moving away from desktop options to online versions. In many ways, this reflects the emergence and underlying benefits of Geoweb technologies, which allow for more distributed GI users and GI contributors. Of course, mobile apps present another suite of issues for institution-wide software deployments that must also be considered. In short, revisitation of relevant use terms is required as higher education increasingly moves to online education and vendors move to products with data plans as apposed to seats or installation limits. As many barriers come down, proprietary software, some scholarship, and many datasets still require pay-for-play to recoup creation costs. A more virtual and centralized deployment from vendors, state agencies, or university consortia may alleviate the need for as many human resources at each institution and GIS access may become as ubiquitous as GI access. The manual tasks from the physical software distribution age may be coming to an end with fewer units, departments, or colleges solely managing site licenses (e.g., DVD delivery and manual installation).

Interestingly, very little data exists on software deployment across U.S. higher education and its institutions. The study detailed above presented a snapshot of software distribution approaches for the products from the largest GIS company in the world. Each individual and the information agency in which they work may have different approaches to administering and distributing software, data, and scholarly communications for their internal uses. Each presents unique challenges and grouping them is problematic in practice given their different intellectual property laws. However, as more open access versions of software, data, and scholarly communications manifest, harmonizing access to them will be essential to advancing knowledge. For example, a researcher may be reading an article on the success of wildlife-crossing structures in national parks, then access the related GI from the article's study, and open that data in GIS software. The technology and policies exist to make this type of action seamless and commonplace, but actions in administration of access to those resources does not quite reflect the current possible levels of openness.

A policy from campus administrators that restricts proliferation of software equates to less work for information providers and reduces costs related to deployment. A more ambitious approach of outreach would increase work for campus administrators, but also the number of GIS-trained students and amount of research using GIS across campuses. Users of GIS include students, faculty, and staff across domains and throughout non-academic campus management activities (e.g., parking). Novice users likely require assistance beyond software access and this will require resources that are typically not included in open source options. However, these are minor impediments. Many GI questions related to geospatial data discovery, fitness for use, and metadata require services beyond the scope of GIS software distribution. Those types of information services exist in academic libraries for other information needs. Unfortunately, rooting the distribution of GIS software and associated information objects in the same location that manages access to scholarly communication is not the norm, despite the potential benefits. Only four academic libraries were identified as managing the Esri site license in this study. It may be useful to learn more about these centralized approaches to software distribution. Twenty-two respondents indicated the site license was administered by IT. Campus IT provides another central and neutral access point to hardware and software on most campuses. This likely adds uniformity to the software suite service across campus for common tools, which should include GIS software. Still, a qualitative study would be required to get at specific processes related to the various approaches employed by IT.

In sum, as GIS tools continue to make inroads to different disciplines it is important to develop strategies that ensure barrier-free access and use to this technology. Consider a world where statistics software would only be available to the faculty, staff, and students of Statistics department who were best trained to use it. Still worse would be the same scenario with word processing and English departments. If those cases were real both the site license study analysis and this book would not exist. It is a ridiculous premise, but GIS software access, at least at some campuses, has not been that far off—offering access to a limited number of expertly trained students and faculty in certain departments (e.g., Geography, Geology, and so forth). This type of limited distribution will no longer suffice. Presently, in reaction to the proliferation of the Geoweb and mobile devices, vendors and campus administrators must open the gates to their GIS software kingdom. Restrictive software distribution strategies unnecessarily limit who receives education and training in GIS and wrongfully dictate how and where GI access and use are negotiated. Fortunately, the market shift from desktop to online software provides easier deployment and removes unnecessary barriers. In a more open, networked, and location-based Geoweb society, gatekeepers maintaining education site licenses must have a mentality that matches the multidisciplinary pursuits in today's higher education. With many GIS-lite tools, vendors have adapted their products and many users continue to demand proprietary software for reasons outlined previously.

Access to GIS software and training is a key factor in decision-making for society. Who has the powerful and valuable skills for GI access and use shape how important decisions will be made. Further study of software deployment of both GIS and other niche tools may provide insights into how to reduce barriers to access and enable new areas of exploration to occur across fields. The same issues of openness depicted in the study of software access surround choices related to the management of GI access and use.

8.4 Conclusion

This chapter focused on human information seeking behavior related to GI access and use. The information seeking behavior of a user to address an information need is informed by their world view, including existing knowledge schemas, terminologies, search strategies, and information avoidance tactics. To study these behaviors and create information services that allow information professionals to help users with their needs, a number of models and theories have been developed. Still, more research needs to be done specifically within the context of GI access and use.

GI in many cases is available, but access and use require work beyond the information organization promulgated through policy and realized by metadata creation. In fact, without assistance from information intermediaries GI access and use for beginners and experts alike can be cumbersome. No amount of information organization or the increasingly incomparable accuracy and precision of modern information retrieval systems will remove the fact that at some point we all need help or do not know how to look for what we need. In information agencies, information services have long been in existence to answer user GI questions, to teach users how to find, evaluate, and use GI on their own, and to create finding aids to simplify GI seeking. In libraries these activities are collectively referred to as reference. This chapter outlined several scenarios where information professionals actively or passively assisted users in locating GI. Finally, because for many GI users, GI access and use requires access to GIS software a review of open source and proprietary GIS software options were discussed along with the findings from one study of software distributions in higher education.

The digital handling of GI, terminology, and new methodologies required the founding of a unique discipline (Goodchild, 1998). Geographic Information Science, as a multidisciplinary field, benefits from the contributions from Information Science related to the study and practice of human information seeking behavior. The work of information professionals to expedite all of the progress made by others through the facilitation of access and use to both GI and software will be indispensable to promote the sharing of knowledge across the fields (Waters, 2007). Chapter 9 presents an overview of the Digital Curation Centre's data lifecycle model to frame all the other work that information professionals do to facilitate GI access and use beyond negotiating GI questions and geospatial technology needs.

References

American Library Association & Reference and Adults Services Division (1996). Guidelines for behavioral performance of reference and information science professionals. *Reference Quarterly, 36*(2), 200–203.

Association of College and Research Libraries (1989). *Presidential Committee on information literacy: Final Report*. Washington, DC: ACRL Retrieved from http://www.ala.org/acrl/publications/whitepapers/presidential.

Barber, J. L. (2002). *Forgotten history: An archaeological perspective on John Sevier at Marble Springs,* Unpublished Master's Thesis (40KN125). Knoxville, TN: University of Tennessee.

Belt, G. T., & Nichols-Belt, T. (2014). *John Sevier: Tennessee's first hero*. Charleston, SC: The History Press.

Berry, B. J. L. (1964). Approaches to regional analysis: A synthesis. *Annals of the Association of American Geographers, 54*(2), 2–11.

Bishop, B. W. (2012). Analysis of reference transactions to inform library applications (apps). *Library & Information Science Research, 34*(4), 265–270. doi:10.1016/j.lisr.2012.06.001.

Bishop, W. W. (1915). The theory of reference work. *Bulletin of the American Library Association, 9*(4), 134–139.

Buckland, M. K. (1991). Information as thing. *Journal of the American Society for Information Science, 42*(5), 351–360.

Case, D. O. (2012). *Looking for information: A survey of research on information seeking, needs and behavior*. Bingley, England: Emerald.

Cassill, D., Tschinkel, W. R., & Vinson, S. B. (2002). Nest complexity, group size and brood rearing in the fire ant, *Solenopsis invicta*. *Insectes Sociaux, 49*(2), 158–163. doi:10.1007/s00040-002-8296-9.

Connaway, L. S., & Radford, M. L. (2011). *Seeking synchronicity: Revelations and recommendations for virtual reference*. Dublin, OH: OCLC Research.

Davie, D. K., Fox, J., & Preece, B. G. (1999). *SPEC Kit 238: The ARL geographic information systems literacy project*. Washington, DC: Association of Research Libraries/Office of Leadership and Management Services, 16.

Dervin, B. (1983). An overview of sense-making research: Concepts, methods, results to date. *ICA presentation*. Dallas, TX.

Dervin, B., Foreman, W. L., & Lauterbach, E. (2003). *Sense-making methodology reader: Selected writings of Brenda Dervin*. Cresskill, NJ: Hampton Press.

Donnelly, F. P. (2010). Evaluating open source GIS for libraries. *Library Hi Tech, 28*(1), 131–151. doi:10.1108/07378831011026742.

Esri. (2015). *History*. Retrieved from http://www.esri.com/about-esri/history

Fisher, K. E., Erdelez, S., & McKechnie, L. (2005). *Theories of information behavior*. Medford, NJ: Information Today.

Gonzalez, A. C., & Westbrock, T. (2010). Reaching out with LibGuides: Establishing a working set of best practices. *Journal of Library Administration, 50*(5–6), 638–656. doi:10.1080/01930826.2010.488941.

Goodchild, M. F. (1998). What next? Reflections from the middle of the growth curve. In T. W. Foresman (Ed.), *The history of geographic information systems* (pp. 369–381). Upper Saddle River, NJ: Prentice-Hall.

Goodchild, M. F. (2003). The nature and value of geographic information. In M. Duckham, M. Goodchild, & M. Worboys (Eds.), *Foundations of geographic information science* (pp. 3–18). New York: Taylor & Francis.

Gross, M. (2002). Integrating the imposed query into the evaluation of reference service: A dichotomous analysis of user ratings. *Library & Information Science Research, 24*(3), 251. doi:10.1016/S0740-8188(02)00125-1.

Janes, J. (2003). What is reference for? *Reference Services Review, 31*(1), 22–25. doi:10.1108/00907320310460852.

MacEachren, A. M. (1979). *How maps work: Representation, visualization, and design*. New York: Guilford Press.

March, G. H. (2011). Surveying campus GIS and GPS users to determine role and level of library services. *Journal of Map & Geography Libraries, 7*(2), 154–183. doi:10.1080/15420353.2011.566838.

March, G. H., & Darnell, B. (2012). Partnering to teach orienteering: The UT Libraries and UT Outdoor Programs' experience. *The Southeastern Librarian, 60*(1), 16–23.

Matthews, J. R. (2007). The evaluation and measurement of library services. Westport, Conn: Libraries Unlimited.

McAuliffe, C. P. (2013). Geoliteracy through aerial photography: Collaborating with K-12 educators to teach the National Geography Standards. *Journal of Map & Geography Libraries, 9*(3), 239–258. doi:10.1080/15420353.2013.817368.

Morville, P., & Callender, J. (2010). *Search patterns*. Sebastopol, CA: O'Reilly Media.

Payne, A., & Singh, V. (2010). Open source software use in libraries. *Library Review, 59*(9), 708–717. doi:10.1108/00242531011087033.

Ramsey, P. (2007). *The state of open source GIS*. Victoria, BC: Refractions Research Inc. Retrieved from www.refractions.net/expertise/whitepapers/opensourcesurvey/survey-open-source-2007-12.pdf

Reference and User Services Association. (2004). *Guidelines for behavioral performance of reference and information service providers*. Retrieved from http://www.ala.org/ala/mgrps/divs/rusa/resources/guidelines/guidelinesbehavioral.cfm

Reiser, C. (2010). *Geographic information systems (GIS) and geospatial analysis global market research study*. Dedham, MA: ARC Advisory Group.

Robinson, A. H. (1952). *The look of maps: An examination of cartographic design*. Madison, WA: University of Wisconsin Press.

Scarletto, E. (2011). Collection development guidance through reference inquiry analysis: A study of map library patrons and their needs. *Journal of Map & Geography Libraries, 7*(2), 124–137. doi:10.1080/15420353.2011.566835.

Shannon, C. E., & Weaver, W. (1949). *A mathematical model of communication*. Urbana, IL: University of Illinois Press.

Smith, M. M. (2008). 21st century readers' aids: Past history and future directions. *Journal of Web Librarianship, 2*(4), 511–523. doi:10.1080/19322900802473886.

Steiniger, S., & Bocher, E. (2009). An overview on current free and open source desktop GIS developments. *International Journal of Geographical Information Science, 23*(10), 1345–1370. doi:10.1080/13658810802634956.

Taylor, R. (1968). Question-negotiation and information seeking in libraries. *College and Research Libraries, 29*, 178–194.

Tobler, W. R. (1970). A computer movie simulating urban growth in the Detroit region. *Economic Geography, 46*, 234–240.

Walford, A. J. (1978). Compiling the "guide to reference material." Journal of Librarianship, 10, 88–96.

Waters, N. (2007). Geographic information systems. *Encyclopedia of library and information science,* (2nd ed., pp. 1106–1115). Taylor & Francis.

Weimer, K. H., Olivares, M., & Bedenbaugh, R. A. (2012). GIS day and web promotion: Retrospective analysis of U.S. ARL libraries' involvement. *Journal of Map & Geography Libraries, 8*(1), 39–57. doi:10.1080/15420353.2011.629402.

Williams, S., Marcello, E., & Klopp, J. M. (2013). Toward open source Kenya: Creating and sharing a GIS database of Nairobi. *Annals of the Association of American Geographers, 94*(1), 37–57.

Wilson, T. D. (1981). On user studies and information needs. *Journal of Documentation, 37*(1), 3–15.

Wilson, T. D. (1997). Information behavior: An interdisciplinary perspectives. *Information Processing and Management, 33*(4), 551–572.

Chapter 9
Data Lifecycle

Abstract This chapter reviews the data lifecycle with respect to geographic information (GI) and details issues related to its digital curation. An overview of data curation profiles, institutional repositories, and spatial data catalogs frame the current approaches and tools used for GI curation. The GI collected and maintained in information agencies continues to change, but each technological advancement in the creation of GI does not exclude the sharing and preservation of older GI types and formats. Therefore, a section on analog collection development and maintenance is included because many information professionals continue to archive print cartographic resources.

9.1 Introduction

Around 67 million years ago, an adult *Tyrannosaurus rex* was having its last meal—an *Edmontosaurus*—then it died, got covered in mud, and slowly fossilized. In 1990, Sue Hendrickson spotted its remains on a commercial fossil hunting trip and the late Cretaceous creature was then saddled with her name, "Sue". The *T. rex*'s remains were found within the boundary of the Cheyenne River Sioux Reservation on land trusted to an individual, but owned by the U.S. Department of the Interior. After a complex round of litigation, fossils were defined as land. In turn, this helped determine U.S. ownership of Sue and the theropod was sold at auction to the Field Museum of Natural History (FMNH) for 7.6 million dollars (Duffy & Lofgren, 1994). These efforts allow visitors to the Field Museum in Chicago to view the specimen with accession number FMNH PR2081, but also enable scientists to study it. For example, a digital scan of its skull allowed scientists to find no evidence to support the popular myth that the left-side skull depression was the result of an attack from another *T. rex* (Brochu, 2003). Similarly, 3D models of other *T. rex's*, compiled through digitization efforts at the Smithsonian Institute, will be available for students to print out by 2019 to facilitate engagement and learning (www.3D.si.edu).

Although this legal saga over fossilized remains is not particularly interesting when compared to the actual life of a *T. rex*, or the education stemming from 3D printouts of bones, the issues associated with its data lifecycle are relevant. Specifically, a careful consideration of the issues centering on data provenance, property and dissemination are critical for advancing science, even if the process

W. Bishop, T.H. Grubesic, *Geographic Information*, Springer Geography,
DOI 10.1007/978-3-319-22789-4_9

takes some time. This was certainly true for specimen FMNH PR2081. But this is also true for complex geospatial data, where developing a deeper understanding of the data lifecycle, data curation profiles, institutional repositories, geospatial data catalogs, and collection development and maintenance will help ensure that data have a long and useful life, helping advance science along the way.

Unlike extinction and evolution in living organisms, there is nothing natural about the selection of data. For example, data that lives another life, beyond its original purpose, proceeds through all the sequential actions of the data lifecycle toward preservation and storage. Most data is allowed to decompose, to deteriorate, or be deaccessioned. For nearly all the data that die, the cause is from the inaction of its creators rather than active deletion. For data that are selected for preservation, its longevity benefits from considerations made at the first step, creation. Subsequently, there are many other important steps along the way, including long-term storage, actions enabling secondary users to find the data, and for the data to be "used with confidence in its authenticity and integrity" (Ray, 2014, p. 2). As detailed throughout this book, both creators of geographic information (GI) and its curators, perform many activities related to GI organization, access, and use. The data lifecycle is an attempt to frame all those activities from the perspective of the data. GI curators keep data alive by: (1) creating or receiving data; (2) appraising and selecting data; (3) ingesting data; (4) preserving data; (5) storing data; (6) accessing, using, and reusing data; and (7) transforming data. All these efforts save the costs associated with data loss and the time it takes to track down and uncover missing data. Anthropomorphizing data in a framework from its perspective is a relatively recent development driven by both federal policy and yeoman labor from the library and archive workforce to improve digital preservation and storage. The following section presents some background on digital curation.

9.2 Digital Curation

For many years, the National Science Foundation (NSF) expected (and assumed) investigators to share data from federally funded projects. There were no formal prescriptive processes or a preservation infrastructure for anyone to follow (Blue Ribbon Task Force on Sustainable Digital Preservation and Access, 2010). In 2011, NSF began requiring data management plans (DMP), and other agencies such as the National Institutes of Health (NIH) and the National Aeronautics and Space Administration (NASA) quickly followed suit. Thus, although formal policies now require compliance from U.S. federally funded projects to ensure data sharing and preservation standardization, all activities related to data worldwide, especially GI, require sharing and preservation to fully reap the benefits from the research investments.

Consider, for example, the complexities associated with measuring interactions between the built environment and the remainder of the biological, geological, and chemical components of the earth, as well as their changes over time. All of this requires digital curation to enable reuse of GI. Examples of decision-making processes and planning efforts that benefit from GI sharing and preservation include but are not limited to "climate change, disaster planning and post-disaster analysis, analysis of land use change and environmental impacts, business and industry site location planning and the resolution of legal challenges" (Morris, 2013, p. 3). Certainly, all decision-making gains from a better understanding and implementation of the data lifecycle. This is especially the case in the data intensive sciences where it is widely recognized that in order for data to be discoverable, accessible, and usable, it must be collected, documented, organized, managed, and curated (Strasser, Cook, Michener, & Budden, 2012). Historically, the research lifecycle concluded with dissemination of findings in a scholarly text but to an "ever-growing extent the published report is accompanied by supplementary data" (Pryor, 2012, p. 6). Traditionally, researchers were incentivized in the production of scholarly publications, but the data were a less disseminated byproduct of the research enterprise. This, however, has changed because of a variety of "open" movements that demand data dissemination to verify results, enable repetition of experiments, and execute new research using the generated data (Higgins, 2012). This swell for data curation followed a larger and earlier movement toward active digital curation of all digital information objects, with advancements such as the creation of tools, organizations, research agendas, and curricula (Higgins, 2011).

Within this context, it is important to note that GI users often lack the human and professional resources to develop meaningful strategies to accomplish the tasks of digital curation. Thus, professional GI curators and their associated expertise provide the resources to mitigate these gaps for many organizations and institutions. This is even more critical for medium and small research projects without the resources to manage data resulting from research activities (Heidorn, 2008). For example, an Inter-university Consortium for Political and Social Research (ICPSR) survey of NSF and NIH funded researchers found only 12% of data produced from awards was archived, and that 45% were shared informally and another 44% were never shared beyond the research team (Pienta, Alter, & Lyle, 2010). This is increasingly the case for smaller projects. GI is at greatest risk for loss or damage "when it is produced by a small group or single person" (Sweetkind-Singer, Larsgaard, & Erwin, 2006, p. 311). Another study of suggested that scientific researchers have a tendency to truncate the data lifecycle and fail to consider steps related to data dissemination, data deposit, data preservation, data discovery, and data repurposing as part of their data lifecycle (Jahnke & Asher, 2012). This is likely the case for most GI creators. Failure to collect details at the time of data creation makes it more difficult for subsequent users to determine the reliability of the data, validate results, and maximize the data's accessibility and reliability to duplicate studies (Higgins, 2012).

"Sharing is at the heart of success, as collecting, storing and making use of data can only come after the means of sharing are in place" (Cragin, Palmer, Carlson, & Witt, 2010, p. 4023). This is where information professionals' contributions to GI organization, access, and use serve as a wrapper for much of the work pertaining to geographic representation. In other words, these curator contributions both *precede* and *follow* the efforts of GI creators to keep the geospatial data alive for others to reuse. For these newer policy requirements, which are driven by sound scientific practices, the dissemination, preservation, storage, and discovery roles for GI have shifted away from the creators to include curators. Data curators embracing this sharing and preservation challenge needed to produce frameworks for the systematic study of the entire data lifecycle. Curators also needed a way to systematically capture the unique aspects of each data type in need of curation. These efforts are detailed in the next several subsections.

9.2.1 Data Lifecycle Model

The activities of GI curators is presented in Fig. 9.1, the UK Digital Curation Centre (DCC) Curation Lifecycle model. The DCC Curation Lifecycle model provides a useful conceptualization of the entire process, as well as activities that enable GI access and use in the present and in perpetuity. Other data lifecycle models exist (e.g., DataONE (https://www.dataone.org/data-life-cycle)), but the DCC data lifecycle model provides a concise overview in a well-designed graphic.

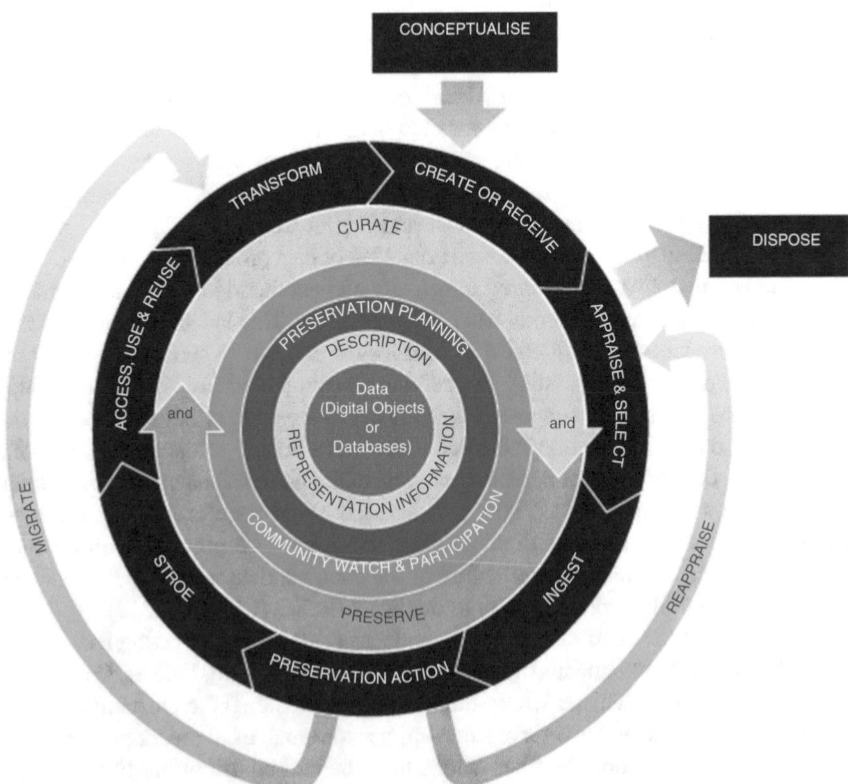

Fig. 9.1 DCC curation lifecycle model

At this juncture, it is appropriate to warn readers of the monotony that lies ahead. Unfortunately, there is no Schoolhouse Rock! video to make this material more palatable, but we will do our best. A review of the model helps congeal future discussion on the work of GI curators. Fortunately, the model standardizes data terminology to enable communication across disciplines conducting curation activities. Digital curators, archivists, and record managers have long received very detailed legally binding instructions on how to manage their GI throughout the lifecycles. The model uses terms that cover broad concepts that permit a range of information professionals to apply the model to their data.

The DCC does acknowledge that the Curation Lifecycle Model is presented in an ideal, sequential flow and "in reality, users of the model may enter at any stage of the lifecycle depending on their current area of need" (http://www.dcc.ac.uk/resources/curation-lifecycle-model). Figure 9.1 rightly suggests that Conceptualization of GI must be completed before any data enters the lifecycle for sharing and preservation. All GI users are familiar with conceptualization and most have experience creating or receiving GI. The center orange circle of the model should also be familiar to all GI users, where the data are the central focus of curation throughout the entire lifecycle. The following provides an introduction to both the full lifecycle actions and sequential actions of this data lifecycle model.

9.2.1.1 Full Lifecycle Actions

The full lifecycle actions reinforce aspects of a data curation mentality and do not relate to tangible tasks. Specifically, these actions are: (1) description and representation information; (2) preservation planning; (3) community watch and participation; and (4) curate and preserve. The activities associated with these actions continue throughout the entire lifecycle (Heidorn, 2011). Description and representation information refer to metadata, which were discussed at length in Chap. 5. The use of appropriate standards in this phase ensures interoperability with administrative, technical and structural, and use metadata, as well as helping users to find, identify, select, and obtain GI through descriptive metadata. Preservation planning (i.e., management) includes overseeing the activities and evaluating the success of all curation objectives. Community watch and participation requires vigilant monitoring of changes in a community of users (e.g., new standards and practices), as well as active engagement with communities to develop those standards, tools, and suitable software. Curate and preserve serves as a final reminder to review the data and to assess the sequential actions throughout the lifecycle.

9.2.1.2 Sequential Lifecycle Actions

The sequential actions of the DCC model provide a linear framework through the stages of the data lifecycle. Beyond conceptualization, the sequential actions include (1) creating or receiving data; (2) appraising and selecting data; (3) ingesting data;

(4) preserving data; (5) storing data; (6) accessing, using, and reusing data; and (7) transforming data. As detailed previously, when a user creates or receives GI, this is an ideal time to capture data provenance related to administrative, descriptive, structural, and technical metadata to determine fitness for use for secondary users. At this point, the creator or curator determines, through appraisal, whether to select GI for curation or dispose of it. Each organization and institution should develop their own procedures to evaluate data and selection criteria for sharing and long-term preservation. Disposal of data occurs for a number of reasons, including compliance with policies or legal reasons. As a result, these policies may call for the secure destruction of the data.

After the appropriate GI is selected for curation, the action of ingestion occurs. This is the point when GI transfers into an institutional repository, spatial data catalog, or other database. Preservation actions are tedious (but necessary), and require the skillset of a digital curator. These actions ensure that data remains authentic, reliable, and usable, by conducting validation checks, quality assurance, quality control. Other chores associated with cleaning data, such as ensuring data structures are in interoperable types and formats, are also completed at this stage. Further, a reappraisal may occur leading curators to reconsider data for re-selection or disposal. Also at the preservation action stage, data migration may occur. This usually requires that the data be transformed to a more usable or sustainable format. The storage stage is simply having the creator or curator make sure that the data is stored securely and in a manner that reduces the chances of data loss (e.g., remote backups).

Access, use, and reuse, are the activities related to discovery. In Higgins (2008), access, use, and reuse are defined as the activities that "ensure that data is accessible to both designated users and reusers, on a day-to-day basis [which] may be in the form of publicly available published information [with] robust access controls and authentication procedures may be applicable" (p. 138). In other words, this stage's actions include all the services and resources used to disseminate, deposit, preserve, discover, and repurpose GI. Again, these information services may be best facilitated by curators because creators may be too busy conducting analysis to answer all the phone calls, emails, or social media pokes from each potential user of their GI. Finally, the transform stage concludes the sequential actions of the data lifecycle. Transforming GI includes actions like converting from original formats to other formats, creating selections or subsets of GI, and deriving new GI by combining various datasets. Many GI creators conduct data transformations frequently, but may not take transformed GI back through the entire data lifecycle. A curator using the sequential stages in the data lifecycle would loop though the steps again to curate and preserve a new transformation. Unfortunately, the knowledge, skills, and abilities required to conduct many of these sequential actions are not part of the typical GI user's toolkit or at least not covered in formal education. This is a critical gap, for many reasons. Namely, curators are not employed in most organizations and institutions creating GI.

9.2.2 Data Curation Profiles

The data lifecycle model provides an overview for others to use when communicating details on stages in the process of making data sharable and preserved. However, each dataset has a unique purpose with a distinctive provenance, ownership, and potentially some restrictions on its distribution. For data intended to be curated, these considerations should be predetermined prior to ingest. The factors related to provenance, property, and dissemination are all documented in data curation profiles (DCPs). A DCP "captures requirements for specific data generated by researchers articulated by the researchers themselves" (http://datacurationprofiles.org/purpose). Creators themselves are not only the best, but likely the only individuals that may provide answers to data questions future users may ask. As a result, the creators serve as the first line of defense in digital preservation (Garrett & Waters, 1996). The creator knows the essential data provenance information to be captured, such as why and the how the data are created, how the data evolves over various stages of the research enterprise (i.e., raw, processed, analyzed, and so forth), and the resulting data types. A DCP allows curators to gather this provenance as well as other reuse information through interview questions that let the creator give a detailed *data story*. The goal of a DCP is to give curators the information needed to best prepare the data, manage its access and use, and design other value-added services, such as enhanced metadata applications and analytical tools, to facilitate sharing and preservation.

With a greater realization that data (both big and small) requires curation, and that curation require training, the role of curator has been filled in many cases by people with backgrounds in librarianship and the other information sciences. GI organization, access, and use covers all aspects of these roles. In the last decade, many academic librarians have explored options in providing data curation services and resources to producers and consumers in higher education. In reaction to the NSF DMP mandate, academic libraries building off of existing information services, such as helping researches locate data, planned to fill the curator role (Fearon, Gunia, Pralle, Lake, & Sallans, 2013). Data type, variety, volume, and forms vary across scientific fields as well as the related information representation practices. This presents challenges to academic libraries filling the curator role because most of the data does not mirror the bibliographic information objects and metadata standards traditionally managed by academic libraries. Although a strong argument has been made for academic libraries to house data on campuses, the DCP may be applied to any organizational entity or individuals tasked with curation.

The DCP Toolkit (2010) provided a methodology, including a data interview, to allow curators to acquire an in-depth understanding of the curation needs of creators and their intended or potential communities of users (Carlson, 2010). Purdue University hosts the DCP Directory (http://docs.lib.purdue.edu/dcp/) with 32 examples ranging from soil science to history. It is highly likely that for any data requiring curation, the creation of a unique DCP and use of the DCP Interviewer's Manual will help gather the provenance, property, and dissemination information in a more systematic way. For example, the curator asks a creator to describe each of the following: (1) an overview of the research; (2) the data, including number of files, size of files, and format of files; (3) the data flow and use in its creation, collection, such as software and metadata; (4) the sharing of data; (5) the depositing of data in an IR; (6) additional documentation need to reuse the data; (7) how to promote discovery; (8) any intellectual property considerations; (9) tools needed to use the data; (10) linking data to publications; (11) measuring the impact of sharing data; (12) backups and updates; and (13) data preservation (Carlson, 2010). Administration of this questionnaire is tedious, and in some instances researchers may not have considered some of these data issues prior to providing data for ingest. For sharing and long-term preservation of GI, these issues must be addressed. Not only does this help determine how best to make GI available for others to reuse, it helps provide confidence in both the quality and legality of use. This first manual also included several questions related to changing the culture of data sharing that may not be necessary for actual ingest of data. Future iterations of the DCP require validation to determine how best to tailor these question for different data types.

The following provides one brief example of how a DCP aides in curation activities. A DCP for fish specimens helps curators provide access and use to those studying ichthyology. Biocollections have historically been kept in university collections (e.g., natural history museums), federal collections, discipline-specific collections (e.g., herbaria), private collections, and zoos, but digital curation of these objects presents new challenges throughout the data lifecycle. The DCP questionnaire asks creators to outline the number of files and file sizes for infrastructure considerations (e.g., images, pdfs of field notes, locality form, loan form, and other ancillary files for each specimen). Preparing a large university's fish collection for sharing and preservation requires the assessment of thousands of files, but for new data from individual researchers to be ingested, the task is less unwieldy because the DCP is designed for ingesting data of a certain scope (and not entire collections) at once. Complicating DCPs, at least in ichthyology, is the common data collection practice of storing multiple specimens of the same species collected at the same time in jars (i.e., lots). These physical lots present users and curators with *collection level access* as opposed to the more desired *item level access* (e.g., fish) for research purposes. Certainly, scientists may still visit museums and open up jars, but the digitization efforts that promise global access to digital biocollections are faced with a unique information access problem for fish data. The lot approach is common knowledge in that research community, but a DCP teases out this aspect of the data itself as a consideration for sharing and preservation.

Biocollections do present many unique curation challenges due to the inherent volume and dynamism of biodiversity. The provenance of fish specimens includes the documentation of location and taxonomy, each with various scientific methods and tools used in the determination of both. Just as the *T. rex* Sue is actually FMNH PR2081, a lot of fish gets a unique identifier to make discovery possible when joining a collection. This unique identifier helps link together, in one record, all the related files for a specimen that may be housed on different servers. For example, one physical lot on a shelf will have a digital text-based record full of metadata and other related digital files (e.g. picture, pdf of field notes, genetic sample). Further, due to the variety of this related data, they require different infrastructures for storage, backup, access, and use. A DCP allows curators to understand the data flow from collection (i.e., fishnet), to curation, including all the software needed to enable reuse (i.e., Internet). Reuse requires creators to divulge more information related to intellectual property and limitations of data reuse. As with Sue, ownership determines who gets to view and manipulate the data downstream and a DCP forces a creator to make those choices. In one study, 60% of participants identified a need to restrict data access for some embargo period until findings were published (Cragin et al., 2010). A DCP helps a curator glean the curation needs of creators and the provenance, property, and dissemination particulars of data. The DCP is another tool that helps fill in gaps of the data's lifecycle related to data dissemination, data deposit, data preservation, data discovery, and data repurposing. Defining these factors for data access and use influence *where* and *how* data should be shared and preserved.

9.3 Institutional Repositories and Spatial Data Catalogs

GI can be housed in a number of ways and the places where data is stored often go by a variety of names. As outlined by U.S. information policy, datasets from Federal Geographic Data Committee (FGDC) members disseminate their GI through a clearinghouse called the GeoPlatform Catalog. Digital metaphors like clearinghouse or warehouse work well for large consortia sharing and preserving GI across many contributors, organizations, and entities. The following presents two common terms for other places that curators manage access and use of data for sharing and preservation.

One option for GI are institutional repositories (IRs), which are often located and affiliated with one institution, but some IRs serve many communities (e.g., ICPSR—the world's largest archive of social science data). One such larger IR, the Dryad Repository (http://datadryad.org), provides many of the services associated with institutional IRs, but for a wider audience. Dryad uses DSpace, a free and open source repository software that preserves and accesses all types of digital content (e.g., text, data, and so forth). Dryad allows scientists to submit data, which may be linked to related publications using that data, and assigns Digital Object Identifiers (DOIs) to enable data citation. These data purposed IRs have a focused scope and audience, but many earlier institutional IRs were created with the simpler task of preserving access to scholarly publications from institution faculty. In some contexts, these archival structures, which were designed to handle text (versus data), mean that many IRs have limited experience curating GI and must overcome significant barriers for achieving success. This includes the technical infrastructure built for other data types and re-training current information professionals (Lyle, Alter, & Green, 2014). Another potential barrier for institutional IRs with GI is the lack of creators providing quality metadata (Markey, Rieh, Jean, Kim, & Yakel,

2007). A recent survey of Academic Research Libraries (ARL), found 88% of those surveyed housed an institutional IR (Fearon et al., 2013). Three-fourths of the academic libraries from that survey use their IRs to share and store data, with others placing data in digital repositories separate from the IR, and still others building data-specific repositories for storage goals. Of note, 84% are absorbing the costs of operating access to data through internal operating budgets and no libraries surveyed are charging for access to the data.

The exponential growth of GI clearly requires some organization to streamline access to all the data and for better or worse, IRs provide one place for creators to deposit data for curation. Academic librarians have long assisted in access and use of GI through the creation of indexes of collections to enable discovery of print cartographic resources in the U.S. (Thiry & Cobb, 2006). Spatial data catalogs serve a similar purpose, allowing users to determine what GI is available and where it is located. A spatial data catalog is an equivalent information organization tool to any online public access catalog (OPAC) for finding books in a library, except it is structured for GI only. This type of list tool may seem antiquated, but for efficiency in discovery, knowing *where* to find *what* saves considerable time, even in a digital domain. To improve access to digital GI, academic librarians help create spatial data catalogs where researchers may deposit their GI. This functions like an IR, but with the added benefit of housing similar data together (Kong, 2015). In fact, efforts have been made to organize spatial data catalogs with similar data across multiple IRs. This does not remove or duplicate data, but provides additional access points to increase its discovery by allowing multiple catalogs to index the same metadata for GI housed in other databases or IRs.

A recent study interviewed several academic libraries to determine the approaches to using spatial data catalogs (Kollen, Dietz, Suh, & Lee, 2013). The results suggest that the most popular was to provide access to digital GI through institution developed spatial data catalogs similar to IRs. For example, the Harvard Geospatial Library provides access to a variety of GI purchased and ready to use as well as manages the curation actions of new GI created by the Center for Geographic Analysis by preserving their data and making it available for other researchers (Guan, Burns, Finkelstein, & Blossom, 2011). Some libraries partnered with local state agencies to provide access to their digital GI (Kollen et al., 2013), while others formed the OpenGeoPortal (http://opengeoportal.org), a multi-institutional partnership of data, to creating and exchanging GI across multiple universities in a consortium. This partnership model resembles earlier efforts to create spatial data catalogs. For example, the National Geospatial Digital Archive project was a partnership between the University of California at Santa Barbara and Stanford University (Erwin & Sweetkind-Singer, 2009). The NGDA provided access and use to considerable digital content of GI from the California Spatial Information Library, the National Map, the National Atlas, as well as scanned data of cartographic resources, such as the David Rumsey Historical Map Collection and the Stanford Geological Survey.

Regardless of the arrangement of these spatial data catalogs, a common feature they shared was the ability for users to spatially browse and spatially search by panning, zooming, dragging, and drawing bounding boxes. This functionality is critical to GI access and use, but is not easily built into IRs focused on information organization and retrieval of text-based objects. GI users benefit from the efforts of all those doing this curation by making GI more easily findable in a variety of formats, connecting to multiple IRs and spatial data catalogs, and increasing the breadth and depth of GI available to them—regardless of where it may be housed. With digital data, this is even more important because many institutions cannot afford nor need to collect or maintain all possible GI. The next section provides a review on some of the strategies and practices to collect and maintain items.

9.4 Collection Development and Maintenance

In 1887, Melville Dewey created the first School of Library Economy at Columbia College (now University) and one of the original six courses concerned accessioning (now Collection Development) (Wiegand, 1999). Although most iSchools or LIS programs no longer require collection development, many still offer this popular elective because information professionals remain responsible for allocating toward the acquisition of information objects to meet their users' information needs. Historically, purchasing referred to the securing of physical materials that built a collection, all of which was housed in a particular location. A collection development plan is the common outcome of such elective courses. For practitioners in any setting, a collection development plan will look like most strategic plans and include a goal, some objectives, steps for the analyses of current collections, and a process for the selection of retention and deaccession (i.e., often ineloquently called *weeding*) of materials. This simple task has grown more complex as many institutions share the management of their digital collections through IRs or spatial data catalogs, or simply have annual access to content through license agreements. Further, significant amounts of GI is now freely accessible through the Geoweb (albeit sometimes not permanently available) (Demas & Miller, 2012). In addition, each object collected (some more than others) requires maintenance beyond its original purchase to ensure preservation.

With GI creation constantly changing, collection development and/or acquisition plans require the flexibility to dynamically adjust to the GI needed, regardless of format (e.g., CD-ROMs at the start of the 1990s for access to journal articles gave way to online databases by the end of the decade) (Larsgaard, 1998). For any information agency's users, the access to their information, as soon as possible in a usable format, is the primary (and sometimes only) concern. GI encompasses many user groups, but the discussion presented here focuses on academic settings with a higher concentration and variety of potential GI users, such as faculty and students, and a history of housing GI.

9.4.1 Collection Development

Collections, whether they include the most haphazard curio cabinet, or a selection of painstakingly cared for comic books, require many of the same tasks performed by information professionals. Acquiring a great deal of information objects is part of the job. However, this comes with the realization that not all things can be collected or all things collected kept. An important caveat is that an information professional is not the end-user. Early archivists warned that archivists should not slip into the assumption that they are the end-user, as this would unintentionally impact the approach to collection, organization, and maintenance (Jenkinson, 1965). This is a difficult habit for some to overcome as the skillset required to be a superb information professional includes seeking, finding, acquiring, organizing, cataloging, and keeping information objects, but real-world restrictions on collection development remove the notion of acquiring or saving everything quickly. Selection of GI for a collection is the result of several pragmatic factors "user demands, budgets, license restrictions, availability, data formats and staffing resources" (Florance, 2006, p. 226).

Objectively determining which GI resources are to be collected is challenging, especially given the dynamic nature of GI, as well as issues of cost, legitimacy, and authority. There are many other resources for GI, but in the U.S., the collection of data provided by the federal government remains unprecedented in volume, authority, and accessibility. For nearly all information agencies working with GI, the pattern of new and available GI long relied on predictable releases from government agencies and commercial firms after laborious data collection through surveys and/or satellites (Larsgaard, 1998). Today, the data availability of most GI is not as predictable, especially given the context of the Geoweb where the data are stored beyond the information agency's control. Some common datasets and subscription databases from vendors purchased in academic settings include SimplyMap, SocialExplorer, LexisNexis U.S. Serial Set Digital Collection, EastView Geospatial and LandScan, World Language Mapping System, and LandInfo Worldwide Mapping LLC. Access to these datasets requires a site license, and although this type of GI is not permanently added to a collection, users may assume a long-standing site license is part of a collection.

A site license to any type of information includes an introduction, definitions, terms and conditions of use, clauses, obligations and responsibilities, and a term of license and termination. An introduction states who the parties to the license are and when agreement becomes effective. Definitions follow which outline each term used in the license (e.g., authorized users, geographic locations, IP ranges, subscription periods, and so forth). The terms and conditions of use state what the product can and cannot be used for. Often the site license outlines multiple subheadings that address acceptable use and prohibited use separately. GI remains information that is expensive to collect, and in academic settings, it is common for educational and research purposes to be acceptable and direct profit to be prohibited. Generalizations for site licenses are problematic as each contract is different. The same product from the same vendor to a similar sized institution typically does not sell at the same rate, but because of non-disclosure clauses, it is difficult to determine how these rates vary.

Other typical clauses address access (secured network, user name and password, and so forth), limits related to circulation of data, copyright provisions, storage, archival access, course packs, course reserves, virtual learning, and sharing between authorized users. The obligations and responsibilities state the expectations of both the *licensor* and *licensee*. The licensor sections address resource delivery, availability of resources, reliability, withdrawal of materials, archival requirements, and usage statistic formats and access to those. The licensee sections include clauses that require steps to be taken to inform authorized users of any usage restrictions, impose a duty to monitor usage, and address breaches of the terms and conditions. The licensee also will need information regarding payments and detailed access information, such as IP ranges. As with most contracts, the term of a license (e.g., start and end) stipulates the period of time the license is in effect, the instances in which the license may be terminated, and who can terminate the license. This section might also outline automatic renewals, price increase caps, and discounts for multi-year purchases.

The headaches of rising costs and shrinking budgets do not necessarily impact licensees symmetrically. For example, we know that data from private vendors can be expensive, but government data may not be. In many ways, government GI is already paid for, but a number of issues related to maintenance cost do result in expenditures. The U.S. government's focus on utilizing information technologies to share GI increased with the passing of the E-Government Act of 2002 (E-Gov) (Tang & Selwood, 2005). As a result of this and other federal mandates, information professionals in academic libraries received less physical GI and the collection development activities shifted from acquiring material to creating guides of hyperlinks, spatial data catalogs, and, in some instances, downloading relevant data into IRs. Today, many geospatial data sets are available through http://catalog.data.gov/dataset or https://www.geoplatform.gov/. Even at the time of this writing, data and its locations are changing and therefore this makes any collection development relying on E-government sources a challenge. Still, allocating human resources to maintain these links to federal, state, and local data commonly accessed by an information agency's users is necessary.

Beyond the vendors and federal, state, and local governments, information professionals may have an obligation to their organization to host multimedia and other digital assets created by local GI creators. Involved academic librarians could take part in teaching, learning, and associated research enterprises at their home institutions (or beyond), in order to have a more central role in assisting users (e.g., faculty, students, and staff) with curation actions. Another large corpus of data that lives and dies on the Geoweb (not related to e-government) may also benefit from data collection. As always, whether volunteered geographic information is useful to the information agencies' users or not is the determining factor for inclusion in collection development plans. As detailed earlier, the current collection development plan for any organization includes rules of selection and deselection of federal and state government documents, copyright considerations, and sustainability of any geospatial data, but these all require revision when local, state, and federal agencies make their data accessible via geoportals. For many, duplicating storage and maintenance of GI will not be possible. However, changes in GI format require a more nuanced understanding of the kinds of resources available from commercial and nonprofit publishers, their avenues of distribution, and new trends in the production and delivery of data. Much of this section on collection development focused on obtaining digital GI, but there remain many publishers selling the other types of GI found in books and atlases, and on globes. Some popular map publishers include International Travel Maps (ITM), National Geographic, Rand McNally, DeLorme, or Omnimap.

9.4.2 Collection Maintenance

Once items are obtained (or not obtained) by an information agency as the result of collection development efforts, several important tasks are required for effective curation. The knowledge, skills, and abilities outlined in the Map and Geospatial Information (MAGIRT) Core Competencies revolved around map scanning and digitization processes, standards, and copyright limitations and materials handling, and preservation methods of both print and digital cartographic items such as encapsulation. The migration from collecting materials towards curating them involves several considerations depending on the object discussed. The maintenance of print and digital materials share some commonalities. To illustrate as many maintenance issues as possible, Sanborn maps will once again be used as an example.

Sanborn maps served as the definitive resource for determining the degree of fire hazard and associated liability in urbanized areas (http://www.loc.gov/collections/sanborn-maps/about-this-collection/). Today, Sanborn maps present a valuable primary resource for historians, urban planners, preservationists, and are often used well beyond their original purpose. When active (1867–2007), the Sanborn Company aggressively extended map coverage to additional cities and provided revisions to existing maps in the form of paste-on correction slips. For the original users, these labels were necessary for accuracy and were pasted over existing maps during the update process. Unfortunately, these paste-on corrections and subsequent updates occurred without any considerations beyond the immediate original purposes. The value of items beyond their original purposes is not predictable, as most GI creators concern themselves with only the most recent version. The largest collection of Sanborn maps and atlases is preserved in the Geography and Map Division, Library of Congress, where there are an estimated 700,000 Sanborn maps in bound and unbound editions. Sanborn serves as a great example of a collection moving from print to digital.

"Paper is matted cellulose fibers held together by the adhesive power of sizing agents such as rosin, starch, and glue" (Larsgaard, 1998, p. 206). For most print maps, the considerations for maintenance of this ephemeral material include housing, arrangement, and preservation. Most Sanborn maps are in massive large bound volumes, but for the purposes of this discussion, imagine one page or one map. Unlike books that are bound and commonly found in stacks, maps lay flat and may be stored in map cases (i.e., flat files or plan files). A few considerations for how maps are placed in cases have long-term impacts, such as flat, folded, and the number of maps to put in a drawer. Filing maps flat in a map case is ideal as paper damages easily and the less wear and tear the better to avoid folds that may lead to tearing. Even if print maps are received folded, the best practice is to unfold them for flat storage (Pritchett, 2013).

The arrangement of print GI must be informed by the community of users. For many information professionals assisting users' information needs with print, the most efficient arrangement is that which helps the information professional—creation of metadata. The creation of metadata for print materials is often referred to as cataloging. Moreover, for physical items, the byproduct of collocation assists users in locating similar GI. Although an alphabetical list might work for smaller print collections, there are benefits to using the Library of Congress schedules to arrange print maps geographically. The Library of Congress G schedule includes separate classification ranges for atlases, globes, and maps. G schedule call numbers are arranged in a geographical hierarchy–starting with the Universe, then the World, then by hemisphere, continent, country, state, counties, cities, and towns. For example, the classification numbers for the University of Tennessee are as follows:

- Regions, etc.: G3962
- Counties: G3963
- Cities and Towns: G3964
- University of Tennessee: G3964.K7:2U5

Maps are fragile by nature and like any other print document, repair may be avoided by not doing certain things. Again, care should be taken to avoid folding maps, using a pen on them, affixing a barcode, or allowing them to circulate. Using a pencil to write the call number works best. If damage does occur, archival tape may be used for tearing. Encapsulation with material such as MylarD makes damaged maps usable and may be easily reversed as none of the MylarD actually binds with the print it protects (Pritchett, 2013). A great deal of tools related to collection maintenance for print GI exist through MAGIRT LibGuide (http://magirt.ala.libguides.com) and the Western Association of Map Libraires' (WAML) Map Librarians' Toolbox (http://www.waml.org/maptools.html).

9.5 Conclusion

All GI users rely on the activities occurring throughout the stages of the data lifecycle to access, use, and reuse geospatial data. The tendency for researchers, GIScientists, and many other GI-related professionals to omit considerations of some stages in the life of their data is understandable. Their interests lie elsewhere, including the spatial and statistical analyses of the data and/or any mapping activities related to GI creation. These jobs do require users to view metadata to determine fitness for use, data provenance, and any limitations or uncertainties that exists, but the other full lifecycle and sequential actions relating to data dissemination, data deposit, data preservation, data discovery, data repurposing, and the activity of data citation do not require attention to accomplish one project, one report, one analysis, one map, or one Geoweb tool. As a result, these tasks may be overlooked in the interest of time. This chapter introduced both a conceptual framework as well as several terms and tasks used by curators to tackle GI organization, access, and use.

Geospatial data without human-driven activities related to GI management and preservation does not live beyond its original purpose and its creator. Data death through discardment, whether intentional or not, is somewhat expected. For example, nearly all the GI from course projects, training activities, and plenty of other exploratory jobs, where more GI is quickly created in just a few clicks (e.g., failed reprojections), are all ephemeral and may be deleted posthaste. Also, the total obliteration of large collections, like that of the museum in Alexandria, are unimaginable in today's digital age, but care still can be taken to share and preserve something (Heller-Roazen, 2002). The entire human record will never be captured and GI is no different. However, considerations made of the entire data lifecycle by those trained to do so does apply to a great deal of GI that are invaluable and irreplaceable. For those GI collections, it is critical to ensure long-term preservation and access. Thus, another role emerges and digital curators exist in Information Science to serve that role by supplementing current practices. With these efforts in mind, the next chapter provides a review and the present educational infrastructure to facilitate these unprecedented preservation efforts. Like the work of a rare book cataloger trained to document every miniscule difference between printings and editions, every GI curator will need to know the minute variances in data and maps, including all GI legacy types and formats, to facilitate GI access and use for others. All GI users benefit from a better understanding of the data lifecycle, data curation profiles, institutional repositories, geospatial data catalogs, and collection development and maintenance, even if they are not ultimately responsible for the curation of the GI they create.

References

Brochu, C. A. (2003). Osteology of Tyrannosaurus rex: insights from a nearly complete skeleton and high-resolution computed tomographic analysis of the skull. *Journal of Vertebrate Paleontology, 22*(4), 1–138. doi:10.1080/02724634.2003.10010947.

Blue Ribbon Task Force on Sustainable Digital Preservation and Access. (2010). *Sustaining the digital investment: Issues and challenges of economically sustainable digital preservation* (Final Report). Washington, DC: National Science Foundation. Retrieved from http://brtf.sdsc.edu/biblio/BRTF_Final_Report.pdf

Carlson, J. (2010). The data curation profiles tookit: Interviewer's manual. *Data Curation Profiles Toolkit.* Paper 2. Retrieved from 10.5703/1288284315651

Cragin, M. H., Palmer, C. L., Carlson, J. R., & Witt, M. (2010). Data sharing, small science and institutional repositories. *Philosophical Transactions of the Royal Society of London*

A: Mathematical, Physical and Engineering Sciences, 368(1926), 4023–4038. doi:10.1098/rsta.2010.0165.

Demas, S., & Miller, M. E. (2012). Rethinking collection management plans: Shaping collective collections for the twenty-first Century. *Collection Management, 37*(3–4), 168–187. doi:10.1080/01462679.2012.685415.

Duffy, P. K., & Lofgren, L. A. (1994). Jurassic farce: A critical analysis of the government's seizure of Sue, a sixty-five-million-year-old Tyrannosauraus rex fossil. *South Dakota Law Review, 39*, 478–528.

Erwin, T., & Sweetkind-Singer, J. (2009). The National Geospatial Digital Archive: A collaborative project to archive geospatial data. *Journal of Map & Geography Libraries, 6*(1), 6–25.

Fearon, D., Gunia, B., Pralle, B. E., Lake, S., & Sallans, A. L. (2013). *ARL Spec Kit 334: Research data management services*. Washington, DC: Association of Research Libraries.

Florance, P. (2006). GIS collection development within an academic library. *Library Trends, 55*(2), 222–235.

Garrett, J., & Waters, D. (1996). *Preserving digital information: Report of the Task Force on Archiving of Digital Information*. Washington, DC: The Commission on Preservation and Access and RLG.

Guan, W. W., Burns, B., Finkelstein, J. L., & Blossom, J. C. (2011). Enabling geographic research across disciplines: Building an institutional infrastructure for geographic analysis at Harvard University. *Journal of Map & Geography Libraries, 7*(1), 36–60.

Heidorn, P. B. (2008). Shedding light on the dark data in the long tail of science. *Library Trends, 57*(2), 280–299.

Heidorn, P. B. (2011). The emerging role of libraries in data curation and e-science. *Journal of Library Administration, 51*(7–8), 662–672. doi:10.1080/01930826.2011.601269.

Heller-Roazen, D. (2002). Tradition's destruction: On the Library of Alexandria. *October*, 100, 133–153. Retrieved from http://www.jstor.org/stable/779096

Higgins, S. (2008). The DCC Curation Lifecycle Model. *International Journal of Digital Curation, 3*(1), 134–140 Retrieved from http://www.ijdc.net/index.php/ijdc/article/viewFile/69/48.

Higgins, S. (2011). Digital curation: The emergence of a new discipline. *International Journal of Digital Curation, 6*(2), 78–88.

Higgins, S. (2012). The lifecycle of data management. In G. Pryor (Ed.), *Managing research data* (pp. 17–45). London: Facet Publishing.

Jahnke, L., & Asher, A. (2012). *The problem of data*. Washington, DC: Council on Library and Information Resources (CLIR).

Jenkinson, H. (1965). *A manuel of archival administration*. London: Percy Lund, Humphries.

Kollen, C., Dietz, C., Suh, J., & Lee, A. (2013). Geospatial data catalogs: Approaches by academic libraries. *Journal of Map & Geography Libraries, 9*(3), 276–295. doi:10.1080/15420353.2013.820161.

Kong, N. (2015). Exploring the best management practices for geospatial data in academic libraries. *Journal of Map and Geography Libraries, 11*(2), 207–225. doi:10.1080/15420353.2015.1043170.

Larsgaard, M. L. (1998). *Map librarianship: An introduction*. Englewood, CO: Libraries Unlimited.

Lyle, J., Alter, G., & Green, A. (2014). Partnering to curate and archive social science data. In J. M. Ray (Ed.), *Research data management: Practical strategies for information professionals* (pp. 203–222). West Lafayette, IN: Purdue University Press.

Markey, K., Rieh, S. Y., Jean, B. S., Kim, J., & Yakel, E. (2007). Census of institutional repositories in the United States. *MIRACLE Project Research Findings*. Washington, DC: Council on Library and Information Resources (CLIR).

Morris, S. (2013). *Issues in the appraisal and selection of geospatial data: An NDSA Report*. NSDA Content Working Group. Retrieved from http://hdl.loc.gov/loc.gdc/lcpub.2013655112.1

Pienta, A. M., Alter, G. C., & Lyle, J. A. (2010). The enduring value of social science research: the use and reuse of primary research data. *The organisation, economics and policy of scientific research workshop*. Retrieved from http://hdl.handle.net/2027.42/78307

Pritchett, H. (2013). *Care and feeding of maps: Tips for managing your map collection* [Webinar]. Map and Geospatial Information Round Table. Retrieved from http://ala.adobeconnect.com/p8v29xakuc6/

Pryor, G. (2012). Why manage research data? In G. Pryor (Ed.), *Managing research data* (pp. 1–16). London: Facet Publishing.

Ray, J. M. (2014). Introduction to research data management. In J. M. Ray (Ed.), *Research data management: Practical strategies for information professionals* (pp. 1–24). West Lafayette, IN: Purdue University Press.

Strasser, C., Cook, R.. Michener, W., & Budden, A. (2012). *Primer on data management: What you always wanted to know*. UC Office of the President: California Digital Library. Retrieved from http://escholarship.org/uc/item/7tf5q7n3

Sweetkind-Singer, J., Larsgaard, M. L., & Erwin, T. (2006). Digital preservation of geospatial data. *Library Trends, 55*(2), 304–314.

Tang, W., & Selwood, J. (2005). *Spatial portals: Gateways to geographic information*. Redlands, CA: ESRI Press.

Thiry, C. J. J., Cobb, D. A., & Map and Geography Round Table (American Library Association) (2006). *Guide to U.S. map resources*. Lanham, MD: Scarecrow Press.

Wiegand, W. A. (1999). Tunnel vision and blind spots: What the past tells us about the present; Reflections on the twentieth-century history of american librarianship. *The Library Quarterly, 69*(1), 1–32.

Chapter 10
Education

Abstract Geographic information (GI) creators, users, and stakeholders exist across nearly all communities, domains, and sectors. Geographic education varies in deployment and delivery for all. Formal training related to GI creation may not emphasize the organization, access, and use aspects of digital curation. Conversely, the existing programs that teach organization, access, and use focus on other information and seldom include coverage of GI. The purpose of this chapter is to outline both the history, current academic landscape, and pave a path forward for educating the different GI-related occupations. We present a multidisciplinary approach that led to the development of one curriculum. The chapter concludes with a call to develop a twenty-first Century GI workforce by coordinating across existing curricular scaffolds from K-12 to graduate programs.

10.1 Geoservices Education

Over the last 20 years, professionals entering fields and sectors reliant on geographic information (GI) have had to adapt their required skillset to the dramatic changes in its creation, dissemination, and reuse. The other chapters in this book outlined issues related to GI organization, access, and use with approaches from the field of Information Science. The opportunities and threats in the education and training for those working in the geoservices also deserve exploration. For now, the opportunities seem endless, with anticipated growth in the workforce and new curricula being developed to prepare individuals for these new jobs. The threats relate to both a lack of understanding of the knowledge, skills, and abilities required to perform these jobs and what the structure should be for GI education and training. Although the areas of geographic information systems (GIS)/geospatial data and digital curation education have largely evolved and operated separately, this chapter explores the area where the two converge and synergistic activities will materialize.

A Google-funded market assessment puts the global revenue for geoservices at $270 billion per year (Oxera Consulting, Ltd, 2013). The rapid growth and expansion in this domain is largely the result of the Geoweb and allied technologies, including open-source data exploitation, crowd-sourcing, distributed computing, and mobile devices. All are changing the way GI is accessed and used, and with it the skill sets required by geographic information professionals (National Research Council, 2013).

W. Bishop, T.H. Grubesic, *Geographic Information*, Springer Geography,
DOI 10.1007/978-3-319-22789-4_10

Within the geospatial field(s), a recent survey showed that mobile and web technologies are likely to gain importance in the next 5 years, along with related topics such as application (app) development (Hofer, Wallentin, Traun, & Strobl, 2014). The "data deluge" described by Hey and Trefethen (2003) has become accepted as a fundamental characteristic of science today, as scientific data continues to increase at a rate of around 30% per year (Pryor, 2015). In this data-intensive science climate, there is a growing recognition that for data to be discoverable, accessible, and usable, it must be collected, documented, organized, managed, and curated (Strasser, Cook, Michener, & Budden, 2012). This led to the National Science Foundation (NSF) requiring data management plans for all proposals in 2011, and other agencies such as the National Institutes of Health (NIH) and the National Aeronautics and Space Administration (NASA) following suit. A great portion of these data are GI. In order to support this rapid growth of geospatial data and jobs, a workforce trained in geospatial technology, GI, and its organization, access, and use is needed.

Due to the diverse, expanding, and interdisciplinary nature of the geoservices, the size, and composition of its workforce can be difficult to determine. The academic competencies to guide future education in the geospatial sciences have received considerable attention from the University Consortium for Geographic Information Science (UCGIS), which has produced several iterations of a *Geographic Information Science and Technology Body of Knowledge* (GISTBoK), funded by NSF (DiBiase et al., 2006, 2010). For Information Science, a number of Library and Information Science (LIS) schools and Information Schools (iSchools) offer curriculum related to digital curation. A matrix of skills and functions for digital curation was one outcome of the *Preserving Access to Our Digital Future: Building an International Digital Curation Curriculum* (DigCCurr I) and *Extending an International Digital Curation Curriculum to Doctoral Students and Practitioners* (DigCCurr II) projects funded by IMLS (Lee, 2009; Poole, Lee, Barnes, & Murillo, 2013). Further, in a review of LIS course catalogs and websites of 55 schools, it was determined that 475 courses in 158 programs contained keywords that appear in the DigCCurr Matrix (Varvel, Bammerlin, & Palmer, 2012). Without exact metrics, we assume that this must be the case for many of the GISTBoK competencies across the course catalogs and curriculum in most Geography departments, as spatial thinking permeates this field. However, in geo-related disciplines and training that focuses solely on the tool—GIS—the metadata competencies may not receive the same coverage. In addition, findings from a recent survey conducted by the Usability & Assessment Working group of DataONE determined that very little data management instruction occurs at the undergraduate level and the most taught topics are Quality Control (21.6%), File Management (20.1%), and Citing Data (19.4%). Yet, these topics are only covered by approximately one-fifth of the 134 instructors surveyed (Tenopir et al., 2015). Predictably one study found that undergraduate students *do not* typically use any standard naming conventions unless instructed to do so for the variety of data they *do* manage (e.g., Microsoft Office, Adobe products, statistical packages, and so forth) (Piorun et al., 2012). The takeaway here is that although the data deluge and Geoweb are relatively new, much more research is needed to know *who* is teaching *what* related to GI organization, access, and use in Geography, Information Science, and other disciplines.

Given these relatively large gaps in our understanding of what is happening with geospatial education in 2016, it is more troubling to realize that researchers noted a serious shortfall in the number of professionals and trained specialists who could work effectively with geospatial data over one decade ago (Gaudet, Annulis, & Carr, 2003). Needless to say, this shortfall in human resources has only grown since then with workforce demands exceeding the output of existing programs (National Research Council, 2015; Prager & Plewe, 2009). In part, this shortfall can be attributed to the inadequate curricula currently offered by many academic departments. Simply put, these units are not providing training quickly or adequately enough at the undergraduate level to fulfill demand (Solem, Cheung, & Schlemper, 2008). The Occupational Outlook Handbook states that employment in fields that use GI data (e.g., Geographic information specialist) are expected to increase at a rate faster than the average rate of growth for all occupations between 2014 and 2024, at a rate of 29% as compared to the average of 7% (U. S. Department of Labor. Bureau of Labor Statistics, 2016). Unfortunately, the Bureau of Labor Statistics (BLS) does not keep data on digital curation as a separate occupation (National Research Council, 2015). Still, several researchers surmise there is a major shortage of data curation professionals (Blake, Stanton, & Saxenian, 2013; Palmer, Thompson, Baker, & Senseney, 2014). One study indicates retirements of baby boomers coupled with the increasing demand for expertise in this area leaves an inadequate supply of librarians, archivists, and curators (Levanon, Colijn, Cheng, & Paterra, 2014). Further, these reports indicate there should be an increase in the overall education in the areas related to GI organization, access, and use.

With changing titles, activities, and expectations, the current training for individuals working in GI organization, access, and use resembles the nineteenth Century model of library education–apprenticing (Wiegand, 1999). Any knowledge-based professional expects lifelong learning on the job; however, there are likely foundational knowledge, skills, and abilities that could provide a head start to those entering this booming profession. This discussion further builds on several key assumptions made throughout the book—namely, that a workforce with a specialization to curate geospatial data and enhance GI access and use is essential to the success of the triple bottom line (i.e., economic, societal, and environmental) in government SDI policies and beyond. As detailed earlier, there is a rich history and infrastructural framework of professionals working in museums, libraries, archives, and data centers that have served as stewards and intermediaries to a rich variety of GI long before it was all digital, but more help and additional infrastructure is required to keep pace with the growth of geospatial data and knowledge. In an effort to succinctly outline the possibilities and constraints of professional education, the next section explores core educational frameworks in both Information Science and Geography to point to a common inception, similar development, and position future collaborations within fields that change with the information they work with.

10.2 Historical Background of GI Creation and Curation

Before 245 B.C.E. when Eratosthenes was the chief librarian of the museum in Alexandria, archivists, librarians, and other information professionals acted as stewards of GI. Today, many within these professions still serve in curator roles—managing recently created GI or that from defunct geospatial technologies. Unlike GI creators and other frequent GI users, these curators amass layers of cartographic and geospatial resources, but are not tasked with creation.

The distinction between creator and curator has roots in the emergence of archival practice and theory in the Netherlands. Today, the implications of such a distinction for professional education remain salient. In post-French Revolution Europe, the purpose of keeping records shifted from the few, to the many. Historically, the goal of archiving was to keep information for *creator-only access*. This was used to preserve the rulers' rights, privileges, ownership, and to serve as "evidence in administrative and legal judgments" (Barritt, 1988, p. 338). To provide Dutch citizens with the ability for more democratic oversight of the operations of a country, archiving approaches changed. Specifically, in 1918, public access to government documents was written into law. In turn, the power associated with unimpeded access to government information was distributed to all citizens through the efforts of the archivists. Best practices for the conscious act of open access stewardship were proposed in *Handleiding voor het Ordenen en Beschrijven van Archieven* (Muller, Feith, & Fruin, 1920). Within this text, two foundational ideals for records management form the ethos for the manual's recommendations—from the French *respect the fonds* (i.e., the original order is kept to reflect the creators' intents, to retain authenticity, and to provide context) and from the German *provenienzprinzip* (i.e., a documented retraceable path to the original source). The enduring influence of respecting the fonds is an ongoing debate for archivists, but the value of provenance remains alive in the modern data lifecycle and data provenance discussions. In fact, it is exponentially more complex, reflecting the ease of duplication, transfer, and transformation of any information object. Of all the important concepts that survived from this early work, perhaps the most important is *that information access is increased when there is a clear division of roles between the curator and the creator*. Whether the impetus is a democracy, sharing amongst scientists, or simply to improve a creators own retrieval ability, the systematic organization of information requires work and this work requires a variety of information professionals.

10.2.1 The Curator

The concepts outlined in this book concerning information organization, access, and use were first formally taught in the U.S. within a storeroom, above the chapel across the street from Columbia University at the School of Library Economy in 1887 (Wiegand, 1996). Within the curriculum for most information professions, echoes from the first school can still be heard when covering content on acquisition, preservation, and utilization of information. In addition, there remains a prevalent stance on pragmatic vocationalism (e.g., service learning), as well as the often cited (but debatable) absence of a theoretical foundation that detaches the discipline from pedagogy and investigative practices found in the rest of social science (Miksa, 1986). For these reasons, library training remained intellectually *across the street* from the educational efforts in other professions, often operating within various large academic and public libraries without any oversight beyond the practitioners teaching in these units (Williamson, 1971).

In 1919, the Carnegie Corporation attempted to professionalize library education, as it had done in medicine, law, and other professions, by funding a study to establish minimum standards of qualification (Williamson, 1971). For information professionals, the lasting implications of the report were the recommendations to restrict library education to those students that had completed a four-year baccalaureate/undergraduate program, to affiliate all schools with a university, and to create a system for accreditation (Williamson, 1923). In turn, the Carnegie Corporation funded the Graduate School at the University of Chicago, which granted the first Ph.D. in LIS, thereby laying a foundation that infused social science research methods to inform library work (Richardson, 1982). An analysis of LIS dissertations from the 1930s to 2000s shows that the largest group of non-LIS chairs are from Education and Theology, which harken back to two aspects of Dewey's school—enculturation to maintain hegemony and undisputed support for education as a public good (Sugimoto, Ni, Russell, & Bychowski, 2011). Prior to the first LIS Ph.D. graduates in the 1930s, committee members needed to come from other seed disciplines.

With the increase in scientific communication at the start of the twentieth Century, documentalists in Europe and the U.S. expanded upon conventional Archival and Library Science techniques to classify, index, and abstract scholarly communication from all the emerging substantive domains and their associated journals in academe (Shera, 1976). Shannon and Weaver's (1949) information theory model observed that information systems are constituted by probabilities (i.e., selection of signals from a well-defined set) and entropy (i.e., disorganization in a system) in a linear, one-way process of signal transmission. The model provided an early definition of information, which was "information is a quantitative measure of freedom of choice available when selecting a message to be sent from the number of possible messages that could be sent" (Raber, 2003, p. 68). This quantification of information matched prior intellectual work in bibliometrics but lacked the universality necessary to account for human complexities of meaning held within information objects. During this era, researchers focused on the signal transfer-type information problems studied in information retrieval (IR) with funding from NSF, NIH, and the U.S. Air Force, amongst others. This work improved the precision and recall of relevant scientific communication in a search—and the backdrop of World War II and the Cold War provided ample incentive to spur advances in the creation of the algorithmic building blocks used for search (and taken for granted by most users), today (Morville & Callender, 2010).

Given the complexity of organizing more types of information objects than books (i.e., journals, data, audio/visual, maps), researchers noticed that a multidisciplinary approach was required to address management issues common across information objects. As a result, a new and inclusive term was coined—*Information Science*. In a presentation, researchers outlined that information science "cannot be equated with documentation, information retrieval, librarianship, or with anything else. Information science is not souped-up information retrieval or librarianship any more than physics is super-charged engineering" (Rees & Saracevic, 1967). As a result, a physical paradigm of information science emerged with methods and theories to study the many new information systems designed to facilitate information organization and access. Library schools with an existing education infrastructure and an adaptable curriculum provided a sensible place within the academy to dock Information Science and faculty poured in from other disciplines. Evidence of this can be found in the Information Science dissertations from the 1970s, where LIS doctoral chairs were from a broader range of disciplines Mathematics, Economics, Political Science all ranked, in order, behind Information Science (Sugimoto et al., 2011).

Rewinding a bit, it is important to note that after years of considerable debate among practicing librarians, professional organizations, and library and information science educators, the 1951 American Library Association (ALA) accreditation standards cemented the master's degree *as the only* degree to be accredited for librarianship. This was done to help increase the status, prestige, and income for the profession of librarianship (Swigger, 2010). This tangentially benefits all other information professionals beyond librarianship. For example, other information professional organizations, such as the Society of American Archivists (SAA), promoted graduate education for workers in their information agencies, but library schools remained the largest home base for information training of all types (Cox, 1988). Another movement for the field was that information scientists began to study the information needs of users (both within and beyond libraries) as opposed to focusing solely on the physical paradigm of bibliography, systems, and IR. For example, Robert Taylor (1968) studied typical reference interviews in library settings and provided a cognitive paradigm for Information Science by wondering *how* and *why* users ask questions, as well as *how* librarians provide an appropriate response to the user's original, often-ambiguous query. This turn toward user-centered design led to a cognitive paradigm that has transformed and informed Information Science education and research as presented in Chap. 8 (Dervin & Nilan, 1986). LIS dissertations from the 1980s into the 2000s, including those with non-LIS chairs and committees, reflect this cognitive trend (Sugimoto et al., 2011). Buckland (1991) reframed the definition of information in the field information science to include physical and digital information objects (i.e., information-as-thing), communication of information (i.e., information-as-process) and cognition (i.e., information-as-knowledge).

In time, many Information schools changed their names by removing information agency-specific terms (e.g., library) to more accurately reflect the educational and research focus of their faculty and associated curricula. The iSchools label and branding effort began in 2005 and today, there are 65 schools in 21 countries that are members of the consortium (http://ischools.org). Many of the values inherited from librarianship remain in these iSchools including "concerns of access, information as a social resource, and the importance of privacy and security" (Dillon, 2012, p. 271). Graduates work in any number of organizations managing information, with a variety of job titles that include Archivist, Bibliographer, Bioinformatics, Cartographic information specialist, Chief Information Officer (CIO), Clinical librarians, Competitive intelligence specialist, Data mining/miner, Database applications specialist/designer, Electronic/digital services specialist/librarian, Geographical Information System Professional (GISP), Health information manager, Indexer, Information architect, Information Resources Manager (IRM), Information scientist, Instructional technology specialist, Intelligence specialist/officer, Legal information specialist/trainer/librarian, Medical informatics, Metadata specialist, Management Information Systems (MIS) director, Museum curator, Ontologist, Preservationist/conservator, Records manager/records and information manager (RIM), Scientific/Technical information specialist, Taxonomist/thesaurus developer, Visual resources specialist, and the classic (and still used), Webmaster.

Although not the case for all, many of these information professionals encounter GI and act as stewards of it. Regardless of what an information intermediary is called, these workers are seldom the creators of the information they manage. Certainly, a great deal of information resulting from the data deluge or other sources does not need curation. The distinction between the creator and the curator is blurred in most cases with those dual roles held by one individual able to produce and share content using Web 2.0 technologies. "Personal information management (PIM) refers to both the practice and the study of the activities people perform in order to acquire, organize, maintain and retrieve information for everyday use" (Jones & Bruce, 2005, p. 2). No one needs or wants assistance organizing their proverbial sock drawer, or collection of music, or memes, but GI organization requires more attention to detail and impacts a larger number of users in most instances than those.

In larger operations, such as the Library of Congress which includes 838 miles of shelves in three buildings, the need for many knowledge workers whose sole tasks are to make information objects organized, accessible, and useable is obvious. This is also the case for school, public, academic, medical, and special libraries of all kinds, as well as museums, archives, and data centers. With more information and data than ever, information professionals that have adapted to new formats (e.g., from analog to digital) have no doubts regarding their job security. Further, it may be best for all other creators of information, especially GI, to reduce their anxiety about big data by acknowledging **two-thirds of GIS is IS**. In short, there is a massive body of knowledge accumulated by an age-old profession with 130 years of experience in U.S. higher education. Help in managing GI can be found here.

The creator and the curator roles differ in purpose (Schellenberg, 1956). For example, creators often will not (and do not) need to consider GI preservation beyond the original purpose of creation. To paraphrase the quote from Goodchild (2013) once more, GI creators or expert users will instantly know the metadata elements by looking at data. Again, the validity of this claim is tenuous, at best. Today, most users of GI, expert or not, require metadata to understand data fitness for use, its provenance, and any limitations or uncertainties that exists. Information professionals are concerned with stewardship of the human record and must take care not to make changes to the original GI or misrepresent any of the GI elements during metadata encoding. These approaches allow for integrity in data provenance and findability beyond the original creator. Although these different worldviews of the same GI are not necessarily competing ones, the two can be at odds with each other because creators may not provide all of the relevant metadata that future users and information professionals need to describe GI. Over time, the lack of detail diminishes the value of GI, its findability, and forever encumbers future use. Each GI professional or individual working in the creation and manipulation of GI *could* take their approach to archiving without training, but this is not a wise decision. A lack of GI education in Information Science leads to inconsistencies in geospatial data management practices (i.e., file naming conventions, authority control, vocabulary control, and so forth) for GI organization. Again, this can impede the discoverability of GI, which is at odds with U.S. federal policy and the overall public sentiment for democratic access and use of information about the world we all share. It also hinders the advancement of scholarship in any number of disciplines that employ GI. Thus, at the very least, GI creators should consult with information professionals prior to curating their data, particularly if they have no formal training in the archival practices.

10.2.2 The Geographer

Chapters 2 and 3 covered important facets of GI creation and its many components. Today, many professionals may need to be both a creator (and an expert user) and an archivist. Consider, for example, a local government GIS analyst. For a project, the analyst may need to acquire the most recent centerlines for all roads in a county. That same analyst likely knows where the centerlines are saved and/or stored within his/her agency. This institutional memory held by one person works well as long as the person remains employed as an analyst. However, when other individuals or analysts work on similar projects, the local conventions for metadata and information organization may not be intuitive (e.g. "usethisone.shp"), especially across agencies. To ensure the seamless and efficient use of data in the future, we must acknowledge that GI creation and GI organization, access, and use are fundamentally different. These differences must inform both practice and education to ensure that a holistic view of data use and its management are adhered to.

In an effort to delineate GI creation (e.g., Geographers) from GI curators, it is important to explore the disciplinary roots of the spatial sciences and their educational framework(s). Within the spatial sciences, there is no doubt that the field of Geography has had the most documented and influential impact on GI education. With apologies, we realize that oversimplification omits continuous contributions from the allied spatial sciences of Geology, Economics, Meteorology, Anthropology, Planning, Public Health, Criminology and many other domains. However, unlike these allied fields, the core mission of Geography is focused on thinking spatially, which is 100% contingent upon GI. With that in mind, the remainder of this subsection provides an abbreviated history of the discipline, focusing on the events that fueled the rise of Geographic Information Science and its educational superstructure. Care is taken to highlight parallels with Information Science.

During Eratosthenes's lifetime, scholars criticized him for having a broad knowledge of the Greek scholarly world instead of expertise in one field (Martin, 2005). Today, Geography retains this breadth of knowledge approach to its study of the world. Given the integrative nature of Geography, it should not be a surprise to readers that the founding father of Geography was a librarian (by trade) and author of *Geographika*, a book that delineated the core paradigms for both physical and human geography. This is not too different from Information Science, where the physical paradigm primarily deals with artifacts and the cognitive paradigm chiefly focuses on users (Ellis, 1992). However, unlike Information Science, Geography tends to focus on one particular type of information object (e.g., GI) or cognition (e.g., wayfinding). Another commonality of Geography with Information Science, in addition to the longstanding debates between the physical and cognitive paradigms, is that unlike the hard sciences (e.g., chemistry, physics), both Geography and Information Science have identity issues (Fenneman, 1919). In both disciplines, there are fears of subsuming meta-disciplines or specialty areas that maintain intentional and inevitable overlap. This is all too obvious in Geography, where the push and pull between the qualitative and quantitative routinely play out in the flagship journal (*Annals of the Association of American Geographers*), and in numerous books and journal articles. These ideological differences and petty disciplinary turf wars are not worth revisiting in this venue (Marcus, 1979; Pickles, 1995). However, it is important to acknowledge that Central Place Theory (Christaller & Baskin, 1966) and its assumptions played a pivotal role in the advent of the Quantitative Revolution in Geography (Schaeffer, 1953), which in turn fueled many of the key methodological developments and epistemology for GIScience. Upon the translations of Christaller and subsequent work by German economist August Lösch (1938), a group of young geographers reading these texts at the University of Washington synthesized these kernels of spatial analysis theory by processing GI with computers and developing many of the methods that form the backbone of spatial analysis today (Openshaw, 1998).

The Quantitative Revolution for Geography began in a department fortunate enough to have achieved two firsts: (1) the University of Washington's Geography department was the first in the U.S. to offer an advanced course in statistical methodology and (2) it was the first unit to acquire an IBM 604 computer (Chrisman, 1998). The results were extraordinary, including the creation of a hierarchical matrix of individual goods, services, and central places (Berry & Garrison, 1958). With these adaptations the Quantitative Revolution helped manifest spatial measurement devices that included "nested hexagons, functional centrality, bid-rent curves, isodapades, trend surface coefficients and computers larger than living rooms" (Barnes, 2004, p. 578). Mastering the mathematical, statistical, and computer-programming techniques required to instruct a computer to perform geostatistical calculations were, at least in part, a foci of the faculty and students. Garrison referred to the group of graduate students as his *space cadets* and their reputation (and myth) grew as many of these students accepted faculty positions in high-profile institutions such as the University of Chicago, Northwestern University, and the University of Michigan (Agnew & Duncan, 2011). In time, these geographers facilitated a shift in curriculum at these institutions, moving from qualitative geographic inquiry to more quantitative topics, including statistical cartography and computerized spatial analysis. This included one of Garrison's former students, Waldo Tobler, who created the First Law of Geography. The first law states, "everything is related to everything else, but near things are more related than distant things" (Tobler, 1970, p. 236). The statement reflects inherent assumptions from Central Place Theory, but it is general enough to reflect all spatial relationships. In fact, Tobler's Law "is at the core of spatial autocorrelation statistics, that is, quantitative techniques for analyzing correlation relative to distance or connectivity relationships" (Miller, 2004, p. 284).

Although it is clear that the Quantitative Revolution in Geography had wide-reaching impacts, the types of citation analysis and related academic genealogy typically performed in Scientometrics and Information Science is largely missing from Geography. This can be resolved, but a great deal of data collection from dissertations across Geography and its subfields would be required to quantify the definitive traces and continued impact of the Quantitative Revolution and GIScience through the discipline. This is not to say, however, that such research is non-existent. For example, in a recent citation analysis study of the term "geographic information systems*" in Web of Science, the six top cited publications for 2003 to 2012 were associated with spatial analysis concepts and methodologies, all of which are rooted in the Quantitative Revolution (Wei, Grubesic, & Bishop, 2015). Further, in looking back to the early 1990s, Goodchild (1992) proposed "Geographic Information Science" as a term that encompasses the scientific questions, methods, and knowledge that transcend the technology of any particular GIS. Interestingly, the coining of Geographic Information Science occurred 25 years after Information Science. The term does codify neatly all the activities that make spatial analysis possible. The volume and variety of these skills complicate education in this evolving area.

Education in the areas of Geographic Information Science are outlined in the GISTBoK under broad headings, including: Analytical Methods; Conceptual Foundations; Cartography and Visualization; Design Aspects; Data Modeling; Data Manipulation; Geocomputation; Geospatial Data; GIS&T and Society; and Organizational and Institutional Aspects (DiBiase et al., 2006). The Geospatial Technology Competency Model (GTCM) was created to define core competencies in a tiered manner, building off of academic and workplace competencies gained in formal education toward industry-wide and industry-specific competencies and appears in Fig. 10.1. The GTCM allows for occupation-specific requirements to be developed through the use of Developing a Curriculum (DACUM) approaches (e.g., GIS Scientist & Technologist) (http://goo.gl/MTMc96).

The sheer scale and scope of the GTCM immediately suggests that most individuals cannot be an expert in all areas. Further, both time and financial restrictions limit what can be learned in formal education programs. GIS education faces five

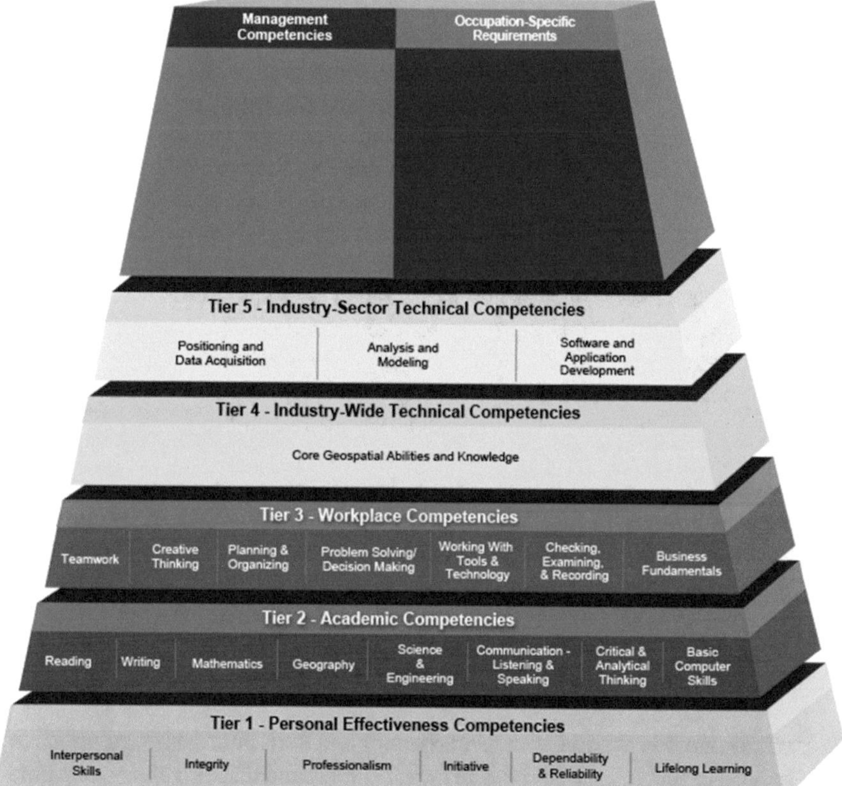

Fig. 10.1 The geospatial technology competency model

other challenges in teaching beyond the multitude of competencies: "1) the incremental nature of GIS subject matter; 2) the element of graphicacy; 3) the need to instill both understanding and skills; 4) the importance of an appreciation of the ontological consequences of adopting a GIS approach; and 5) the need to develop students' project-management skills for real-world applications" (Whyatt, Clark, & Davies, 2011, p. 236). The rapid advances in a growing industry make aligning curricula with employer needs difficult; however, given the importance of spatial literacy in everyday life, K-12 education has been charged with teaching GIS across the curriculum (National Academies Press, US, 2006). Further, because GIS skills are in such demand from employers, other parts of the academy and the private sector are driving GI education in reaction to an unmet demand (Bearman, Munday, & McAvoy, 2015). In 2013, for example, the U.S. Geospatial Intelligence Foundation developed accreditation guidelines for the Collegiate Geospatial Intelligence Certificate to fill the authority void by providing external review and quality assurance for academic programs in GI.

A third player worth noting here, beyond the creator and the curator, is the tool developer. Just as information professionals' roles focus on storing and retrieving GI created by others, GI creators are typically employed in the application and/or marketing of GIS rather than any actual system development (Longley, 2000). Although some information scientists and some geographers play roles in system design, the majority do not play any role in creating the proprietary systems they use. Future work proposed on usability would allow for some feedback on existing systems from current end-users. However, education in this arena is in its infancy, and it is difficult for most to keep pace with the rapid development of these tools. As detailed by Sui (2004), interdisciplinary thought will spark creative solutions to many GIS problems, but this is proving to take some time. That said, the GI organization, access, and use problems could benefit immensely from the many lenses of Information Science and its obvious overlap with Geographic Information Science. Interestingly, map librarians have been at the forefront of this union space between the two disciplines for many years. In fact, it is the work of map librarians that led to a federally funded project that was the impetus for this book. Therefore, map librarianship provides a first look at a multidisciplinary approach to curriculum development with implications for all creators and curators.

10.3 Geographic Information Librarianship

In 1937, Walter Ristow, the Map Librarian of New York Public Library, conducted the first assessment of map libraries in the U.S. He found that only thirty map libraries existed in the U.S. with at least one full-time employee. Further, only 25% of these libraries were academic in nature. However, after World War II, map libraries roles changed quickly and began playing a crucial role in national defense efforts by cataloging massive quantities of printed maps from Europe and Asia, ensuring that the U.S. would never be cartographically unprepared for future conflicts (Murphy, 1969). In some ways, this made many academic librarians the most influential information professionals in the U.S., as they developed strategies for managing and organizing this influx of cartographic resources vital to national security.

Specifically, the Army Mapping Service (AMS) (now the National Geospatial-Intelligence Agency), selected forty-five academic libraries to begin acquisitioning thousands of maps as part of the Federal Depository Library Program, a program that provides government documents "for public use without charge" (Nicoletti, 1971; Wilhite, 2011). As a result, many public university map libraries serve not only faculty, students, and staff, but also provide open access to their archived materials to any member of the general public. Throughout the twentieth century, many more academic libraries acquired other maps from the Government Printing Office (GPO) and some still receive print maps to this day from the Department of the Interior's Minerals Management Service and the Central Intelligence Agency (CIA). In addition to government materials, academic, public, and special map libraries retain collections of unique (often local) print cartographic resources (e.g., maps). Because many of these maps and data are rare and/or unavailable elsewhere, many map librarians must retain the skill of cataloging to make print cartographic resources discoverable. Keeping these skills is important because not all libraries have the technology, human resources, or budget to preserve and/or digitize these materials for online access (Hansen, 2012). Further, as the growth of digital resources exploded both within the private sector and federal, state, and local governments in the U.S., all map libraries were forced to alter collection development, acquisition, and archiving plans to include CD-ROMs and other digital media formats in the 1990s (Larsgaard, 1998).

As alluded to previously, almost all information professionals working in libraries, archives, data centers, museums, and other information agencies face management challenges for print items. Unfortunately, where the management of digital resources are concerned, the supporting literature and research from allied areas are somewhat sparse. Although many of the tenure-track librarians are incentivized to publish and share best practices through the academic literature, this is not true for other professionals outside of academia. As a result, it is likely that many clever management strategies were never (and still have not) been disclosed. In an effort to mitigate these knowledge gaps within academic libraries, the Association of Research Libraries (ARL) *SPEC Kit 219 Geographic Information Systems* provided a suite of best practices, including the types of GIS services already offered at different academic libraries (Soete & Adler, 1997). In turn, the ARL Geographic Information Systems Literacy Project built upon the first SPEC kit and found that 89% of the seventy-two responding academic libraries surveyed provided GIS services. It is important to remember that prior to the Geoweb, this was the only type of digital GI services available (Davie, Fox, & Barbara, 1999). These data spawned a number of practice-oriented studies that explained the different GIS-related services developed and/or found in academic libraries (Abresch, Hanson, Heron, & Reehling, 2008; Aufuth, 2006; Houser, 2006; Johnston & Jensen, 2009; Kowal, 2002; Strasser, 1998). As a result, geographic information librarians (GIL) emerged as a term to define those librarians that facilitate the collection, dissemination, and use of both digital GI as well as traditional analog cartographic resources (Weimer & Reehling, 2006).

Simultaneously, there are related developments within other types of information agencies, including the emergence of digital archivists, museum curators, and many others working to wrangle the digital deluge of GI. For example, the Library of Congress funded the National Geospatial Digital Archive project to collect, preserve, and provide long-term access to at-risk geospatial data that includes more than ten terabytes of geospatial data and imagery (Erwin, Sweetkind-Singer, & Larsgaard, 2009). Another U.S. project, the Geospatial Multistate Archive and Preservation Partnership (GeoMAPP), collected snapshots of GI that were often at risk of being overwritten by updates and changes to real-time geospatial data (Morris, 2013). Also, eScience professionals and their data curation roles and required skills to management new data types were explored (Kim, Addom, & Stanton, 2011). In turn, similar partnerships are beginning to emerge between academic libraries and government agencies in the digital curation of GI (Hoover, 2012). The knowledge, skills, and abilities across these different information agencies to provide these services are likely very similar, especially given the analogous purposes. Education beyond apprenticing informed by research will help to make GI organized, accessible, and findable for as many potential users as possible. Aside from a smattering of map cataloging courses in LIS programs and the continuing education events from the ALA Map and Geography Round Table (MAGERT) Education Committee, there is not formal education in this area.

10.3.1 The Geographic Information Librarianship Project

In 2012, the Institute of Museum and Library Services (IMLS) funded the Geographic Information Librarianship project (2012–2014) through the Laura Bush twenty-first Century Librarian Program to bolster the data curation education of librarians and improve their abilities to deal with the varied resources generated by geospatial tools. This was a collaborative effort between Drexel University (Tony H. Grubesic, PI) and the University of Tennessee (Bradley Wade Bishop, PI). The proposal was crafted to highlight the great need for information professionals educated in GI and serve the growing number of GI users across commerce, government, the military, and academia. In academic libraries alone, fifty jobs were advertised in academic libraries from 2007 to 2009 (Mandel & Weimer, 2010). Similar growth continues in academic library positions with twenty-seven unique positions advertised in 2015 on the popular listserv Maps-L (https://listserv.uga.edu/archives/maps-l.html). In response to the occupational sea change, the MAGERT (established in 1980 and the largest map library organization in the world) (http://www.ala.org/magirt/front) voted to change its name to the Map and Geospatial Information Round Table (MAGIRT) in 2011. MAGIRT members were strong supporters of this project. The following outlines the curriculum development and evaluation aspects of the *GIL* project as an example for future educational work.

10.3.1.1 MAGIRT Core Competencies

In the context of this massive growth and change in GI, the Education Committee created the 2008 core competencies (CCs). The CCs were specific to librarians specializing in GI and under three specializations: (1) Map Librarianship; (2) GIS Librarianship; and (3) Map Cataloging and Metadata Creation (MAGERT, 2008). These competencies were created in response to a charge from the ALA Executive Board's Task Force on Library Education to create "actionable recommendations to ensure that library and information science education programs produce librarians who understand the core values of our profession and possess the Core Competences of the profession needed to work in today's libraries" (American Library Association, 2009, p. 1). ALA Round Tables and Divisions were simultaneously asked to create their core competences.

The intention of the CCs went beyond their initial LIS education program reform task, but hoped "to assist in the professional development life cycle: from student/faculty curriculum development to new professional to mid-career professionals or others who are new to the specialization, as well as administrators or personnel officers to assist in job descriptions and hiring in this area" (MAGERT, 2008, p. 2). The development of the 2008 CCs included the following steps: (1) twelve members of the Education Committee drafted CCs list; (2) a draft circulated to members for feedback; (3) suggested changes were incorporated into a final document (MAGERT, 2008). The document reflected the three distinct type of librarians that emerged from typical job descriptions at that time. However, since 2008 most new GIL jobs have required overlapping knowledge, skills, and abilities—in effect—fusing these once distinct career tracks. From a pragmatic perspective, this fusion of skills and abilities helps meet the growing demand for GI services. Further, in a world where the budgets for all information agencies continue to shrink, it is also indicative of the need for versatile professionals that have the capacity to manage both print and digital cartographic resources, metadata, and provide cataloging services. In total, 75 CCs were identified—suggesting that these GILs were expected to have a near encyclopedic expanse of knowledge, skills, and abilities. This proves to be an educational challenge, especially given the constraints of iSchool infrastructure, which is built atop the foundation of LIS and largely reflects courses dominated by bibliographic and other text-based information objects and content. In a market with many other GI education options (http://www.geotechcenter.org/geospatial-national-map.html), what role do iSchools and LIS programs play (if any)?

10.3.1.2 Survey Validation

Despite this extreme growth in GI and related jobs, the 65 American Library Association (ALA) accredited LIS programs, and most iSchools do not have educational opportunities for their students to learn these skills. There are a handful of programs that offer a stray geospatial technology course related to learning desktop software or Geoweb tools, but the curriculum for GI is relatively thin, at best, in most of these LIS and iSchools. This is unfortunate because there is strength (and expertise) embedded within iSchools and LIS programs that could focus on GI organization, access, and use. That said, many of the current information professionals working with GI learned on the job or came from other areas, including various GIS backgrounds. To address the dearth of coursework, one outcome of the *GIL* project was the development of two courses. These courses did not focus on GI creation or spatial analysis, but instead on the **IS in GIS**.

To give footing to what the IS in GIS might be, the CCs from MAGIRT were used. These courses did not downplay all the other knowledge, skills, and abilities that are extremely useful when working with GI, but the scope of items covered are more appropriately taught in other disciplines' GIS classes (e.g., geographic and cartographic competencies). These *GIL* courses fixated on the most important aspects of information professional work given the constraints of limited time. One way it accomplished this was to strip the geography and cartography content to roughly one-third of the total course material. In an effort to facilitate this reduction and focus on the most vital topics for GI organization, access, and use, a survey validation of working professionals was used.

The survey validation approach provided a feedback mechanism, which allowed information professionals beyond MAGIRT Education Committee members to rank the most important CCs from their everyday, real-world information agency experience. This process was geared toward creating both a timely and relevant curriculum and the necessary data to revise future iterations of the MAGIRT CCs. The survey validation approach is commonly used by a variety of other professions to infuse empirical data from job incumbents into coursework (Raymond, 2005). An unavoidable limitation of using the CCs was their Education Committee origin. Certainly, the ideal start to constructing CCs would involve more advanced job analyses (e.g., observation) with a stratified sample of workers. Due to potential biases from those practitioners heavily involved in professional service with an interest in GIL education, the current CCs required validation from others working in the area to reduce biases.

The survey included occupational questions about each participant's job and educational background. In addition, participants were asked to rank CCs on a scale from 0-Not Important to 3-Very Important or select "I do not perform this task". After input from the Advisory Committee and university institutional review boards (IRB) approval, a web-based survey was launched and promoted through recruitment e-mails to the lists of several relevant organizations and working groups (MAPS-L, a listserv for map and geographic information librarians; magirt@ala.org, the listserv for all MAGIRT members; the Western Association of Map Librarians listserv; the Association of Canadian Map Libraries and Archives (ACMLA) listserv; the GIS 4 Librarian; GOVDOC-L, a listserv for government documents librarians; North American Cartographic Information Society (NACIS) listserv). An additional 42 personal emails were sent to archivists from two National Digital Information Infrastructure and Preservation Program (NDIIPP)–funded projects.

One-hundred and fifty-seven information professionals responded to the survey and the greatest number of respondents reported their job title as "map librarian" (37). Other job titles included GIS librarian (21), map cataloger (13), archivist/records manager (11), manager (16), and several positions with disciplinary terms from the spatial sciences that were followed by "librarian" (e.g., Geography, Urban Planning, Earth Science, Geology, and so forth). Respondents were also asked where they worked and the majority of survey respondents (76%) defined their work setting as either a main or a branch/special academic library. Other work settings included map libraries, education units teaching GIS, state government, and private for-profit libraries. Nearly 45% reported holding a master's in LIS. The only comparable study ever was conducted in 1976 when Ristrow found only 20% held a master's in LIS. It is safe to assume the percentage increased steadily from 1976, but current findings show there are still many individuals working with GI in information agencies without education in how GI is organized, accessed, and used. As outlined throughout this book, a focus on metadata creation, end users' search, and other behaviors are critical pieces to prepare anyone assisting with stewardship of GI. Perhaps this lack of LIS training reflects the fact that those trained to work with GI are not a product of ALA-accredited master's degree programs without GI courses. Twenty-nine individuals held master's degrees in Geography, Geology, Urban Planning, or other GIS-related fields, 35 held bachelor's degrees in those disciplines, and 6 had subject area PhDs. Other degree areas mentioned by participants were Geomatics and Information Technology. The variety of educational backgrounds reflects the many avenues that lead to working with GI.

The CCs importance ratings ranged from 1.45 to 2.62, with an average standard deviation of 0.75 on a 0 to 3 point scale, with an option to select "I do not perform this task" and with those already having relatively low rated importance means were removed from further consideration in the curriculum development. By drawing the cutoff of importance at 2.3, 23 CCs remained, and those were rewritten as thirteen student learning outcomes for the GI courses in Table 10.1.

The most important CCs were supplemented by Knowledge Areas of Cartography and Visualization (CV) and Geospatial Data (GD) from the GISTBoK. Again, the GISTBoK provides educators with knowledge areas, units, topics, and learning objectives that are, "applicable across the undergraduate, graduate, and postbacca-

Table 10.1 GIL Student Learning Outcomes

Student learning outcomes	
Course section	Student learning outcome
1. Geography and cartography	*1.1 Students will demonstrate geographic and cartographic principles, including geographic and cartographic scale, projection, grids, and geographic coordinate systems*
2. Collection development / Records Appraisal/ collection maintenance	*2.1 Students will demonstrate knowledge of local, state/provincial, federal and international mapping agencies and private map publishers, map series and similar publication patterns, and gazetteers, data portals, volunteered geographic information, and aspects of the Federal Depository Library Program* *2.2 Students will select strategies to obtain different types of maps, imagery, and other geospatial data* *2.3 Students will describe copyright considerations and the ability to negotiate licensing agreements for databases and collections of geographic information* *2.4 Students will explain how to assess the strengths and specialties in a collection and the needs of users to inform collection development* *2.5 Students will describe proper materials handling, especially for rare and fragile materials*
3. Reference and instruction	*3.1 Students will demonstrate the ability to locate geospatial data and software support* *3.2 Students will gain awareness of GIS tutorials & training* *3.3 Students will develop and deliver geographic information consultations*
4. Metadata/ cataloging	*4.1 Students will explain metadata standards, schemas, and issues* *4.2 Students will understand and interpret existing metadata in geospatial records* *4.3 Students will define projections, coordinate systems, and other physical characteristics of cartographic items to create metadata records* *4.4 Students will interpret and calculate cartographic scale*

laurate/professional sectors of GIS&T education infrastructure" (DiBiase et al., 2006). Combined these two educational tools provided a multidisciplinary approach to GI curriculum development informed by practice and may assist others initial steps. More detailed discussion of the project method and implications exists in Bishop, Cadle, & Grubesic (2015).

10.3.1.3 Student Success from First Offerings and Future Iterations of the Courses

The students in both sections of the first GI courses at the University of Tennessee and Drexel University participated in pretest and posttest assessment based on these student learning outcomes. The average scores improved by 13.1% (Bishop, Grubesic, & Parrish, 2015). To look for a significant difference in this change, a multivariate test of the results for both programs was done and found a ($F(1,23) = 20.913$, $p < .001$). The improvements in the (1.) Geography and Cartography ($p < .001$) and (4.) Metadata/Cataloging ($p < .003$) sections of the test were statistically significant for students at both universities. Unlike non-LIS programs, students likely gained knowledge on (2.) Collection Development /Records Appraisal/ Collection Maintenance and (3.) Reference and Instruction in other coursework as these tactics and practices transfer regardless of the type of information object. A full account of the study appears in Bishop, Grubesic, & Parrish (2015). If the test were given to non-LIS students, results may show that students with geographic and cartographic competencies would improve most in the last three (more IS) sections. Although finding and evaluating GI, handling and managing GI, and creating metadata for GI are likely valuable skills to anyone working in the geoservices, without testing it is not known how well these competencies are known. Without close review of other curriculums, it is also not known if these competencies are taught outside the iSchools.

The survey validation approach to validate CCs, the use of CCs to develop student learning outcomes, and the student learning outcome assessment met the educational objectives of the GIL project. These same three tasks provide future educators a systematic approach to curriculum development in emerging areas, broadly speaking. Sharing the approaches as well as specific materials, lectures, and tests will benefit education in these areas to increase standardization and improve quality of content of IS in GIS. At the University of Tennessee, the *GIL* courses are taught biannually as 543: Geographic Information in Information Sciences and 516: Introduction to Geospatial Technologies. The courses, and the MAGIRT CCs, require future validation work and revision because many aspects of policy, metadata, resources, users, and technology change. This book is the final outcome directly attributed to the *GIL* project and serves as an effort to reach beyond iSchools to share the knowledge, skills, and abilities useful across the geoservices. After all, it is increasingly clear that many GI creators require education of basic archiving practices, metadata, and many other facets of GI that are important to data management plans, critical to data reuse, and understanding the provenance of data acquired for use (Mayernik et al., 2015).

10.4 Evidence-based Geoservices Education

Another effective strategy for developing geoservices education relies upon a more grassroots approach. For example, in a more regularized educational environment, where oversight is important, the use of the GISTBoK or GTCM would be ideal for all programs. Programs would exhibit little deviation from the evidence-based knowledge, skills, and abilities that current occupations demand. However, both the GISTBoK and GTCM struggle, over time, because of the rapid advances in geospatial technologies, information overload, the changing demands of an expanding workforce and the varied and growing amount of users (Prager, 2012). Additionally, the GISTBoK was criticized for focusing on lower cognitive skills (DeMers, 2009). One solution to mitigate this problem is to extend the lower skills to K-12 education. Although the GTCM purports to give a framework for specializations in the geoservices to fill-in those missing pieces by using DACUMs, there is much work left to be done. These obstacles to evidenced-based education in the geoservices are not insurmountable, and the *status quo* of educators relying on their work backgrounds and research areas lacks coordination. The spontaneity of course design bringing real-world experiences and cutting-edge knowledge to the classroom should be fostered, as the expert philosopher is a huge asset in higher education for teaching people to *think* and not just *do*. Evidenced-based geoservices education will not alter the thinkers approaches to teaching, but it will provide an intellectual foundation grounded in workforce trends to produce employable "doers" for the twenty-first Century.

10.4.1 Developing a Curriculum (DACUM) and K-12 Outreach

A "Developing a Curriculum" (DACUM) is a job analysis technique that uses job incumbents to develop core competencies. The ALA and MAGIRT CC development examples outlined in this chapter were done with both task forces and committees. It is fair to say that these experts may have been the most qualified to determine the talent required for work in their fields and specializations. A utilitarian approach also exists throughout librarianship's history of practice, but for the construction of education rooted in current needs, the committee of experts approach is not representative of the many others working in the area that are not highly involved in professional service.

The DACUM method offers advantages over other job analyses because of a relatively lower cost and time commitment (e.g., direct observation). In order to clearly explain DACUM, a few definitions are required. Job incumbent refers to professionals doing the work at present and these professionals are rarely leaders or educators in their field. Administration and management generally includes individuals that performed similar jobs in the past, but being removed from day-to-day operations, their involvement introduces inaccuracies in the representation of current work. This differentiation between roles is important when considering the information professions and geoservices, especially given the number of external changes in technology and information policy. With these definitions in mind, a DACUM is based on three core principles. First, job incumbents know their job better than anyone else, and therefore they are the best at describing what it is they do. Second, the best way to define a job is by describing the *specific tasks* that are performed on the job. Like any other intellectual work, those employed in information agencies routinely have tasks that may be difficult to describe. Regardless, the professionals who are currently performing those tasks are the best able to explain clearly what those tasks are. Finally, the third principle of the DACUM is that all tasks performed on a job require the use of knowledge, skills, and abilities (KSA) that enable successful performance of those tasks. This may not be a given for all professions, but success is one justification for education.

In its most basic form, the DACUM process consists of a two-day workshop that features a trained DACUM facilitator that leads a discussion with five to twelve job incumbents, referred to as Subject Matter Experts (SMEs). It may be prudent to hold DACUMs in conjunction with professional meetings. However, when possible, this should be avoided in an effort to eliminate bias and sample evenly across a profession. In addition, "the panel facilitator should be an individual who has had no experience with the profession" to reduce biases (Knapp & Knapp, 1995). A strong DACUM facilitator also ensures that all participants have an equal voice in the process. Upon completion of the initial workshop, the facilitator will draft a DACUM chart, which is a document that provides a linear representation of all of the tasks performed by job incumbents along with all enablers required to perform each task. In addition to the linear representation of the job, a DACUM chart includes lists of knowledge, skills, and abilities required of job incumbents and may include other lists related to the profession (i.e., lists of resources and technology used, industry terminology). A DACUM chart is then circulated to a broader representation of the profession through a survey to validate the most important and critical competencies to include as central topics in coursework. Ideally, a DACUM approach would be used across the geoservices and indeed is proposed GCTM's next steps to integrate actual work into classroom discussions, assignments, and lectures in undergraduate and graduate education.

These higher education elements of the infrastructure are key, but K-12 must be included. Outreach materials using DACUM results could also be crafted to share with K-12 teachers to teach elements of the CCs from *GIL*, data science, and Geographic Information Science, to early learners. This would give students the lower cognitive skills necessary to succeed in a world where spatial and information literacy are so valuable. As stated in the recent National Research Council report, *Preparing the workforce for digital curation*, an educational continuum is required that "will include graduate-level education in digital curation for some, discrete study programs and certificates for others, perhaps supplementary courses inserted into established curricula in other fields, or exposure through online courses and conferences" (NRC, 2015, p. 63). In other words, education on digital curation everywhere for everyone and GI will permeate all curricula in this area as long as DACUM findings reach all those audiences.

10.5 Conclusion

The chapter provided a detailed history, the current status, and a need for further development in GI organization, access, and use education. The promise of jobs combined with the multidisciplinary positioning of geospatial technologies results in a large number of educators involved in training the twenty-first Century geoservices workforce. Who teaches what should be a secondary concern to what is taught. The *GIL* project and subsequent DACUM and outreach proposal both provide some approaches to building education in the different GI-related occupations. Further job analyses will also be required to better understand the different occupations working with GI and to produce relevant education and training over time.

Like any educational effort, teaching methods should be structured to facilitate and optimize learning, regardless of the subject. If current practitioners around the GI-related professions created competencies in a systematic way, students would benefit the most. Finally, requiring the evaluation of student learning outcomes at the course and program levels and developing new assessments will add accountability for educators and ensure student success. The singular takeaway from this chapter should be that education in this area (no matter who teaches it) should blend both the Geographic Information Science and Information Science knowledge, skills, and abilities needed for successful GI creators and GI curators because both roles need to be done well to move this domain and its constituents forward.

References

Abresch, J., Hanson, A., Heron, S., & Reehling, P. (2008). *Integrating geographic information systems into library services: A guide for academic libraries*. Hershey, NY: Information Science Publishing.

Agnew, J. A., & Duncan, J. S. (2011). *The Wiley-Blackwell companion to human geography*. New York: Wiley.

American Library Association. (2009). *American Library Association's Core Competences of Librarianship*. Retrieved from http://www.ala.org/educationcareers/careers/corecomp/corecompetences

Aufuth, J. (2006). Centralized vs. decentralized systems: Academic models of GIS and remote sensing activities on campus. *Library Trends, 55*(2), 340–348.

Barnes, T. J. (2004). Placing ideas: Genius loci, heterotopia and geography's quantitative revolution. *Progress in Human Geography, 28*(5), 565–595.

Barritt, M. (1988). Archival training in the land of Muller, Feith, and Fruin: The Dutch national archives school. *The American Archivist, 51*(3), 336–344.

Bearman, N., Munday, P., & McAvoy, D. (2015). Teaching GIS outside of geography: A case study in the School of International Development, University of East Anglia. *Journal of Geography in Higher Education, 39*(2), 237–244. doi:10.1080/03098265.2015.1010146.

Berry, B. J. L., & Garrison, W. L. (1958). A note on Central Place Theory and the range of a good. *Economic Geography, 34*(4), 304–311.

Bishop, B. W., Cadle, A. W., & Grubesic, A. (2015a). Job analyses of emerging information professions. *Library Quarterly, 85*(1), 64–84.

Bishop, B. W., Grubesic, T. H., & Parrish, T. (2015b). Mapping LIS elective across the field: Collaborative student learning outcome development and assessment. *Journal of Education for Library and Information Science, 56*(4), 272–282.

Blake, C., Stanton, J. M., & Saxenian, A. (2013). Filling the workforce gap in data science and data analytics. *iConference 2013 Proceedings* (pp. 1015–1016). Retrieved from https://www.ideals.illinois.edu/bitstream/handle/2142/42501/424.pdf?sequence=4

Buckland, M. K. (1991). Information as thing. *Journal of the American Society for Information Science, 42*(5), 351–360.

Chrisman, N. R. (1998). Academic origins of GIS. In T. W. Foresman (Ed.), *The history of geographic information systems: Perspectives from the pioneers* (pp. 33–43). Upper Saddle River, NJ: Prentice Hall.

Christaller, W., & Baskin, C. W. (1966). *Central places in southern Germany*. Englewood Cliffs, NJ: Prentice-Hall.

Cox, R. J. (1988). Educating archivists: Speculations on the past, present, and future. *Journal of the American Society for Information Science, 39*(5), 340–343. doi:10.1002/(SICI)1097-4571(198809)39:5<340::AID-ASI11>3.0.CO;2-0.

Davie, D. K., Fox, J., & Barbara, P. (1999). *SPEC Kit 238: The ARL Geographic Information Systems Literacy Project*. Washington, DC: Association of Research Libraries, Office of Leadership and Management Services.

DeMers, M. N. (2009). *Fundamentals of geographical information systems*. Minnesota: Wiley.

Dervin, B., & Nilan, M. (1986). Information needs and uses. *Annual Review of Information Science and Technology, 21*, 3–33.

DiBiase, D., Corbin, T., Fox, T., Francica, J., Green, K., & Jackson, J. (2010). The new geospatial technology competency model: Bringing workforce needs into focus. *URISA Journal, 22*(2), 55.

DiBiase, D., DeMers, M., Johnson, A., Kemp, K., Luck, A. T., Plewe, B., et al. (2006). *Geographic information science and technology body of knowledge*. Washington, DC: Association of American Geographers.

Dillon, A. (2012). What it means to be an iSchool. *Journal of Education in Library and Information Science, 53*(4), 267–327.

Ellis, D. (1992). The physical and cognitive paradigms in information retrieval research. *Journal of Documentation, 48*(1), 45–64.

Erwin, T., Sweetkind-Singer, J., & Larsgaard, M. L. (2009). The national geospatial digital archives—Collection development: Lessons learned. *Library Trends, 57*(3), 490–515.

Fenneman, N. M. (1919). The circumference of geography. Geographical Review, 7(3), 168–175. http://doi.org/10.2307/207825.

Gaudet, C. H., Annulis, H. M., & Carr, J. C. (2003). Building the geospatial workforce. *Urisa Journal, 15*(1), 21–30.

Goodchild, M. (2013). *Personal communication*. April 4, 2013.

Goodchild, M. F. (1992). Geographical information science. *International Journal Geographical Information Systems, 6*(1), 31–45.

Hansen, C. M. (2012). Saving the Ratzer Map: Lessons learned in the conservation, restoration, management, and publicity of cartographic resources. *Journal of Map & Geography Libraries, 8*(3), 264–275. doi:10.1080/15420353.2012.700913.

Hey, T., & Trefethen, A. (2003). The data deluge: An e-Science perspective. In F. Berman, G. Fox, & T. Hey (Eds.), *Grid computing: Making the global infrastructure a reality* (pp. 809–824). Chichester, England: Wiley.

Hofer, B., Wallentin, G., Traun, C., & Strobl, J. (2014). *Workforce demand assessment to shape future GI-education—First results of a survey*. 17th AGILE Conference on Geographic Information Science. Castellón, Spain. Retrieved from https://agile-online.org/Conference_Paper/cds/agile_2014/agile2014_145.pdf

Hoover, J. (2012). GIS Collaborations in Saskatchewan: SGIC and the University of Saskatchewan Library. *Journal of Map & Geography Libraries, 8*(1), 68–79. doi:10.1080/15420353.2011.622601.

Houser, R. (2006). Building a library GIS service from the ground up. *Library Trends, 55*(2), 315–326.

Johnston, L. R., & Jensen, K. L. (2009). Maphappy: A user-centered interface to library map collection via a Google maps "Mashup". *Journal of Map & Geography Libraries, 5*(2), 114–130. doi:10.1080/15420350903001138.

Jones, W., & Bruce, H. A. (2005). *A report on the NSF PIM workshop*. Retrieved from: http://pim.ischool.washington.edu/final%20PIM%20report.pdf.

Kim, Y., Addom, B. K., & Stanton, J. M. (2011). Education for e-Science professionals: Integrating data curation and cyberinfrastructure. *The International Journal of Digital Curation, 6*(1), 125–138.

Knapp, J. E., & Knapp, L. G. (1995). Practice analysis: Building the foundation for validity. In J. C. Impara & L. L. Murphy. (Eds.), *Licensure testing: Purposes, procedures, and practices*. Lincoln, NE: University of Nebraska Press.

Kowal, K. C. (2002). Tapping the web for GIS and mapping technologies: For all levels of libraries and users. *Information Technology and Libraries, 21*(3), 109–114.

Larsgaard, M. L. (1998). *Map Librarianship: An introduction*. Englewood, CO: Libraries Unlimited.

Lee, C. (2009). *Functions and skills (Dimension 2 of matrix of digital curation knowledge and competencies)*. Retrieved from http://www.ils.unc.edu/digccurr/digccurr-functions.html.

Levanon, G., Colijn, B., Cheng, B., & Paterra, M. (2014). *From not enough jobs to not enough workers: CFO implications*. Ottawa, ON: The Conference Board of Canada Retrieved from: http://www.conferenceboard.ca/e-library/abstract.aspx?did=6489.

Longley, P. A. (2000). The academic success of GIS in geography: Problems and prospects. *Journal of Geographical Systems, 2*(1), 37–42.

Lösch, A. (1938). The nature of economic regions. *Southern Economic Journal, 5*(1), 71–78.

MAGERT. (2008). *Map, GIS and Cataloging/Metadata Librarian Core Compentencies*.

Mandel, L. H., & Weimer, K. H. (2010). Necessary skills for map and GIS librarians: Job descriptions for inform LIS curriculum. *Association for Library and Information Science Education*. Boston, MA.

Marcus, M. G. (1979). Coming full circle: Physical geography in the twentieth century. *Annals of the Association of American Geographers, 69*(4), 521–532 Retrieved from http://www.jstor.org/stable/2563126.

Martin, D. (2005). Socioeconomic geoComputation and e-social science. *Transactions in GIS, 9*(1), 1–3. doi:10.111/j.1467-9671.2005.00201.x

Mayernik, M. S., Thompson, C. A., Williams, V., Allard, S., Palmer, C. L., & Tenopir, C. (2015). Enriching education with exemplars in practice: Iterative development of data curation internships. *International Journal of Digital Curation, 10*(1), 123–134.

Miksa, F. L. (1986). Melvil Dewey: The professional educator and his heirs. *Library Trends, 34*(3), 359–381.

Miller, H. J. (2004). Tobler's first law and spatial analysis. *Annals of the Association of American Geographers, 94*(2), 284–289.

Morris, S. (2013). *Issues in the appraisal and selection of geospatial data: An NDSA Report*. Washington, DC: NSDA Content Working Group Retrieved from http://hdl.loc.gov/loc.gdc/lcpub.2013655112.1.

Morville, P., & Callender, J. (2010). *Search patterns*. Sebastopol, CA: O'Reilly Media.

Muller, S., Feith, J. A., & Fruin, R. (1920). *Handleiding voor het Ordenen en Beschrijven van Archieven* (2nd ed.). Groningen, The Netherlands: Erven B. Van der Kamp.

Murphy, M. (1969). History of the army map service map collection. In R. W. Stephenson (Ed.), *Federal Government MapCollecting: A brief history* (p. 3). Washington, DC: Special Libraries Association.

National Academies Press (US) (2006). *Learning to think spatially*. Washington, DC: National Academies Press.

National Research Council (2013). *Future U.S. workforce for geospatial intelligence*. Washington, DC: The National Academies Press.

National Research Council (2015). *Preparing the workforce for digital curation*. Washington, DC: The National Academies Press.

Nicoletti, F. T. (1971). U.S. Army Topographic Command College Depository Program. *Special Libraries Association Geography and Map Division Bulletin, 86*, 2–3.

Openshaw, S. (1998). Towards a more computationally minded scientific human geography. *Environment and Planning A, 30*(2), 317–332.

Oxera Consulting, Ltd. (2013). *What is the economic impact of geo services?* Retrieved from http://www.oxera.com/Oxera/media/Oxera/downloads/reports/What-is-the-economic-impact--of-Geo-services_1.pdf?ext=.pdf.

Palmer, C. L., Thompson, C. A., Baker, K. S., & Senseney, M. (2014). Meeting data workforce needs: Indicators based on recent data curation placements.

Pickles, J. (1995). *Ground truth: The social implications of geographic information systems.* New York: Guilford Press.

Piorun, M. E., Kafel, D., Leger-Hornby, T., Najafi, S., Martin, E. R., Colombo, P., et al. (2012). Teaching research data management: An undergraduate/graduate curriculum. *Journal of eScience Librarianship, 1*(1), 8. doi:10.7191/jeslib.2012.1003.

Poole, A. H., Lee, C. A., Barnes, H. L., & Murillo, A. P. (2013). *Digital curation preparation: A survey of contributors to international professional, educational, and research venues. Association for Library and Information Science Education.* WA: Seattle.

Prager, S. D. (2012). Using the GIS&T Body of Knowledge for curriculum design: Different design for different contexts. In D. Unwin, K. Foote, & N. Tate (Eds.), *Teaching geographic information science and technology in higher education* (pp. 63–80). Chichester, West Sussex: Wiley Blackwell.

Prager, S. D., & Plewe, B. (2009). Assessment and evaluation of GIScience curriculum using the geographic information science and technology body of knowledge. *Journal of Geography in Higher Education, 33*(S1), 46–69. doi:10.1080/03098260903034012.

Pryor, G. (2015). Why manage research data? In G. Pryor (Ed.), *Managing research data* (pp. 17–45). London: Facet Publishing.

Raber, D. (2003). Librarians as organic intellectuals: A Gramscian approach to blind spots and tunnel visions. *Library Quarterly, 73*(1), 33–53.

Raymond, M. R. (2005). An NCME instructional module on developing and administering practice analysis questionnaires. *Educational Measurement: Issues and Practice, 24*(2), 29–42.

Rees, A. M., & Saracevic, T. (1967). Towards the identification and control of variables in information retrieval experimentation. *Journal of Documentation, 23*(1), 7–19.

Richardson, J. V. (1982). *The spirit of inquiry: The Graduate Library School at Chicago, 1921–51.* Chicago: American Library Association.

Schaeffer, F. K. (1953). Exceptionalism in geography: a methodological examination. *Annals of the Association of American Geographers, 43*(3), 226–249.

Schellenberg, T. R. (1956). *Modern Archives: Principles and Techniques.* Chicago: The Union of Chicago Press.

Shannon, C. E., & Weaver, W. (1949). *The mathematical theory of communication.* Murray Hill, NJ: Lucent Technologies.

Shera, J. H. (1976). Two centuries of american librarianship. *Bulletin of the American Society for Information Science, 2*(8), 39–40.

Soete, G. J., & Adler, P. S. (1997). *Issues and innovations in geographic information systems* (Vol. 219). Washington, DC: Association of Research Libraries.

Solem, M., Cheung, I., & Schlemper, M. B. (2008). Skills in professional geography: An assessment of workforce needs and expectations. *The Professional Geographer, 60*(3), 356–373. doi:10.1080/00330120802013620.

Strasser, T. C. (1998). Geographic information systems and the New York State Library: Mapping new pathways for library service. *Library Hi Tech,16*(3), 43–50.

Strasser, C., Cook, R., Michener, W., & Budden, A. (2012). *Primer on data management: What you always wanted to know.* California Digital Library: UC Office of the President Retrieved from: http://escholarship.org/uc/item/7tf5q7n3.

Sugimoto, C. R., Ni, C., Russell, T. G., & Bychowski, B. (2011). Academic genealogy as an indicator of interdisciplinarity: An examination of dissertation networks in Library and Information Science. *Journal of the American Society for Information Science and Technology, 62*(9), 1808–1828. doi:10.1002/asi.21568.

Sui, D. Z. (2004). GIS, cartography, and the "third culture": Geographic imaginations in the computer age. *The Professional Geographer, 56*(1), 1–157.

Swigger, B. K. (2010). *The MLS project: An assessment after sixty years*. Lanham, MD: Scarecrow Press.

Taylor, R. (1968). Question-negotiation and information seeking in libraries. *College and Research Libraries, 29*, 178–194.

Tenopir, C., Allard, S., Sinha, P., Pollock, D., Newman, J., Dalton, B., Frame, M., & Baird, L. (in press, 2016). Data management education from the perspective of science educators. *International Journal of Digital Curation*.

Tenopir, C., Allard, S., Sinha, P., Pollock, D., Birch, B., Dalton, B., Frame, M., & Baird, L. (2015). *Data management education from the perspective of science educators*.

Tobler, W. R. (1970). A computer movie simulating urban growth in the Detroit region. *Economic Geography, 46*, 234–240.

U. S. Department of Labor. Bureau of Labor Statistics. (2016). *Occupational Outlook Handbook, 2015–16*. Retrieved from: http://www.bls.gov/ooh/.

Varvel, V. E., Bammerlin, E. J., & Palmer, C. L. (2012). *Education for data professionals: A study of current courses and programs*. In Proceedings of the 2012 iConference (pp. 10–12). doi:10.1145/2132176.2132275

Wei, F., Grubesic, T. H., & Bishop, B. W. (2015). Exploring the GIS knowledge domain using CiteSpace. *The Professional Geographer, 67*(3), 374–384. doi:10.1080/00330124.2014.983588.

Weimer, K. H., & Reehling, P. (2006). A new model of geographic information Librarianship: Description, curriculum and program proposal. *Journal of Education for Library and Information Science, 47*(4), 291–302. doi:10.2307/40323822.

Whyatt, D., Clark, G., & Davies, G. (2011). Teaching geographical information systems in geography degrees: A critical reassessment of vocationalism. *Journal of Geography in Higher Education, 35*(2), 233–244. doi:10.1080/03098265.2010.524198.

Wiegand, W. A. (1996). *Irrepressible reformer: A biography of Melvil Dewey*. Chicago: American Library Association.

Wiegand, W. A. (1999). Tunnel vision and blind spots: What the past tells us about the present; reflections on the twentieth-century history of american librarianship. *The Library Quarterly, 69*(1), 1–32.

Wilhite, J. M. (2011). *The evolution of public printing in the United States*. Retrieved from: http://libraries.ou.edu/locations/docs/govdocs/federal.ppt.

Williamson, C. C. (1971). *The Williamson reports of 1921 and 1923, including Training for library work (1921) and Training for library service (1923)*. Metuchen, NJ: Scarecrow Press.

Index

W. Bishop, T.H. Grubesic, *Geographic Information*, Springer Geography,
DOI 10.1007/978-3-319-22789-4

Printed in the United States
By Bookmasters